Tanks: 1914-1918

Lieut. Colonel Sir Albert G. Stern, K. B. E.

Tanks: 1914-1918
The development of Allied Tanks and Armoured Warfare during the Great War.

Albert G. Stern

Tanks: 1914-1918
The development of Allied Tanks and
Armoured Warfare during the Great War.
by Albert G. Stern

First published under the title
Tanks: 1914-1918

Leonaur is an imprint
of Oakpast Ltd

Copyright in this form © 2010 Oakpast Ltd
Text translated to modern English

ISBN: 978-0-85706-143-0 (hardcover)
ISBN: 978-0-85706-144-7 (softcover)

http://www.leonaur.com

Publisher's Notes

In the interests of authenticity, the spellings, grammar and place names used have been retained from the original editions.

The opinions of the authors represent a view of events in which he was a participant related from his own perspective, as such the text is relevant as an historical document.

The views expressed in this book are not necessarily those of the publisher.

Contents

The Idea	9
Experiments	19
Trials—and Trials	33
First Orders	53
Tanks in Action	77
Production on a Large Scale	91
Fighting the War Office	118
The War Office Gets Its Way	145
The Anglo-American Treaty	150
The Tanks Get Their Way	165
In Conclusion	194
Appendix 1	199
Appendix 2	201
Appendix 3	221
Appendix 4	222
Appendix 5	223
Appendix 6	231

TO MY STAFF

CHAPTER 1

The Idea

AUGUST 1914 TO FEBRUARY 1915

My experiences in the great war may be of interest to a peaceful world in years to come.

In July 1914 rumours of war brought up threatening clouds in a sky already darkened by strikes and revolutionary unrest among the industrial workers.

Banking circles in London were already more than cautious, and when the crash came we had anxious moments. I remember well attending a meeting of bankers at the Bank of England, under the chairmanship of Lord Cunliffe. Sir Edward Holden was the commanding figure. "I must pay my wages on Friday," he said, "and we must have Bank Holidays until enough currency has been printed to be able to do so." His advice was followed and all the impending disasters were averted.

Here, I first saw (in the war) the advantage of a definite cure administered by a strong man.

At this time, also, I saw Mr. Lloyd George at work. It was on a question of Bills of Exchange.

We had meetings of a committee under the chairmanship of Mr. Huth Jackson. Mr. Lloyd George, as Chancellor of the Exchequer, had issued a proclamation which would not meet the case. It was a matter of life and death to a great many in the City. Mr. Jackson had an interview with Mr. Lloyd George one evening. Before he could report to our committee the next day, a new proclamation had been published making the necessary alterations.

Here, again, I learnt the advantage of a strong man ready to act and take responsibility in the sudden-changing conditions of war.

My two brothers were both in the Yeomanry, one a major in the South Irish Horse, the other a lieutenant in the Westminster Dragoons. I had a broken ankle which had always given me much trouble and prevented me from volunteering in the ordinary way. I meant, however, to offer my services to the State in some fighting capacity so soon as we had tackled the many difficult problems which arose in the city.

In November I wrote to Mr. Churchill, First Lord of the Admiralty, offering to provide and equip an armoured car, with crew complete, to be attached to Lieut. Spencer Grey's bombing squadron in Flanders, and received the following reply:—

Admiralty,
Whitehall,
December 6th, 1914.

Dear Sir,
Mr. Churchill asks me to say that he has read your letter telling him that you have offered to equip and command, at your own expense, an armoured car for Squadron-Commander Spencer Grey's aeroplane unit.

In reply the First Lord wishes me to say that in his opinion an armoured car would be of little use to this unit, and that it would be much better if you were to arrange an interview with Captain Sueter, the Director of the Air Department at the Admiralty, and offer the services of the car and of yourself to the regular Armoured Car section which is being built up and organised under Commander Boothby at Wormwood Scrubbs.

The First Lord much appreciates your generosity in making the offer, and feels sure that in doing as he suggests you would be putting it to the most effective use.

Yours faithfully,
J. Masterton Smith.

I saw Captain Sueter, who was in charge of the Armoured Car Division of the R.N.A.S., at the Admiralty, and was given a commission as Lieutenant, R.N.V.R. Commander Boothby was our C.O., and Major Hetherington Transport officer to the division.

Major Hetherington asked me to join his staff and to work under his Chief Assistant, Lieut. Fairer-Smith. I agreed to do so.

We were stationed at Wormwood Scrubbs, in the *Daily Mail* Airship Shed, and were in process of forming armoured car squadrons and motorcycle machine-gun squadrons; there was no accommodation in our camp and no roads, but plenty of mud. The Airship Shed was our only building.

Enthusiasm was great; the personnel was full of rare talent, engineering and otherwise, and there was nothing that wanted doing but we could find an expert among us to do it. Discipline was rather conspicuous by its absence. Squadrons grew at a great rate, fully equipped, and, as an American actor used to sing, they were *"all dressed up and no place to go."*

Mr. Churchill had hoped, as at Antwerp, to be able to play an important rôle on the Belgian coast, co-operating his own land forces with the Fleet; but the Army did not appreciate Mr. Churchill's naval land forces, and although he had a perfect right to send them to Dunkirk, which was under the direction of the Admiralty, they were never very welcome in the zone of the armies.

The Duke of Westminster took the first of these squadrons to Flanders, and other squadrons quickly followed, but even the Duke's tact and position and the excellent quality of the squadron under him could not overcome the difficulty of being under the Admiralty while serving with the army. Still earlier than this. armoured cars, under Commander Samson, had operated during and after the evacuation of Antwerp from the racecourse at Ostend.

Our fine naval land force was well equipped, well armed, and had a splendid personnel, but with all the fighting on land at the disposal of the War Office and not of the Admiralty, its officers found that they could only get scope for it by their own personal efforts, by using their imaginations and the influence of their friends. Lieut.-Commander Boothby got his squadron and Lieut.-Commander Josiah Wedgwood's squadron to the Dardanelles. Boothby, Commander of Squadron 1, one of our best officers, and certainly one of the most delightful, capable and unassuming of men, was killed on landing. Josiah Wedgwood, the most fearless of men and a great leader, was very severely wounded on landing.

Lieut.-Commander Whittall took a squadron to German West Africa, where they were very much handicapped by the terrain, but they helped to win a battle and were thanked by both Houses of Parliament of the Union. This squadron returned and, I believe, went later to British East Africa. Ford cars, on account of their lightness and high clearance, were most useful in this campaign.

The Duke of Westminster later took a squadron to Western Egypt, with great success against the Senussi. Commander Locker Lampson got a squadron to Russia, and all the world has heard of its exploits, operating from the frozen seas, in Russia, in Poland, and as far south as the Caucasus.

The enthusiasm of both officers and men of the Armoured Car Division was unbounded. They searched the whole world for war. But war in France had already settled down to trench fighting. In France, armoured cars, always an opportunist force, found all opportunities gone.

Major Hetherington, our transport officer, had distinguished himself in the early days of airships. He was young and always full of new ideas. He had a great knowledge of motorcars, although not an engineer, and there was no new invention which he would not eagerly take up and push forward.

After discussion among certain officers and civilians about the uselessness of armoured cars, except on roads, and the great strides that had been made in light armour plate as protection against the German "S" bullet Major Hetherington got the Duke of Westminster sufficiently interested in the idea of a landship to invite Mr. Winston Churchill to dinner.

Already, before this, at a supper of three at Murray's Club, Hetherington, James Radley and myself, a proposal had been put forward to build a landship with three wheels, each as big as the Great Wheel at Earl's Court. In those days we thought only of crossing the Rhine, and this seemed a solution.

I also remember Hetherington proposing to fire shells at Cologne by having a shell which, when it reached the top of its trajectory, would release a second shell inside it, with planes attached, and this second shell would plane down, making 100 miles in all. It is strange that the Germans later tried and succeeded in firing about eighty miles, but not in this way.

Mr. Churchill came to the dinner and was delighted with the idea of a cross-country car. Commodore Sueter, Lieut.-Commander Briggs and Major Hetherington made the following suggestion:—

> It may be briefly described as a cross-country Armoured Car of high offensive power.
>
> It consists essentially of a platform mounted on three wheels, of which the front two are drivers and the stern wheel for steering, armed with three turrets, each containing two 4-inch

ROLLS-ROYCE ARMOURED CAR

ARMOURED LORRY WITH 3-POUNDER GUN

Sir Eustace Tennyson D'Eyncourt, K.C.B.

guns, propelled by an 800h.p. Sunbeam diesel set, electric drive to the wheels being employed. The engines, as well as the guns and magazines, would be armoured, but not the purely structural part, which would be fairly proof against damage by shell fire if a good factor of safety were used and superfluity of parts provided in the structure.

The problem of design has been cursorily examined by Air Department officers, and the following rough data obtained:—

Armament	3 twin 4-inch turrets with 300 rounds per gun.
Horse Power	800, with 24 hours' fuel, or more if desired.
Total weight	300 tons.
Armour	3-inch.
Diameter of wheels	40 feet.
Tread of main wheels	13 feet 4 inches.
Tread of steering-wheel	5 feet.
Over-all length	100 feet.
Over-all width	80 feet.
Overfill height	46 feet.
Clear height under body	17 feet.
Top speed on good country road	8 miles per hour.
Top speed on bad country road	4 miles per hour.

The above particulars must be regarded as approximate and cannot be guaranteed, owing to the absence, in the department, of technical knowledge properly applicable to this problem. These particulars are, however, quoted in the belief that they can be readily worked to.

The cross-country qualities of the machine would appear to be good. It would not be bogged on any ground passable by cavalry. It could pass over deep water obstacles having good banks up to 20 feet or 30 feet width of waterway. It could ford waterways with good bottom if the water is not more than 15 feet deep. It could negotiate isolated obstacles up to 20 feet high. Small obstacles such as banks, ditches, bridges, trenches, wire-entanglements (electrified or not) it would roll over easily. It could progress on bottom gear through woodland of ordinary calibre.

The greatest disabilities of the machine appear to be as follows:—
It cannot cross considerable rivers except at practicable fords, which practically means that it cannot operate as a detached unit in country held by the enemy, where this involves the systematically opposed crossing of big rivers. It can be destroyed by sufficiently powerful artillery. It can be destroyed by land mines.

The machines might on occasion do good service by destroying railway lines in the enemy's rear, but its most important function would appear to be in destroying the enemy's resistance over any region where he does not possess other guns than field guns or howitzers.

It would appear at first sight that the machine ought to be more heavily armed and gunned, but considerations of the disproportionate weight of the guns and of the time of building have resulted in the proposal being reduced to the comparatively moderate one described above.

This was much the same fantastic idea that Mr. H. G. Wells had developed in one of his stories years before.

Mr. Churchill then set up a committee to study the question, and Mr. Eustace Tennyson d'Eyncourt, C.B., the Director of Naval Construction, was appointed chairman on the 24th of February, 1915. It was to be known as the Landship Committee.

Before this date Mr. Churchill had already written his now historic letter to Mr. Asquith:—

My Dear Prime Minister,
I entirely agree with Colonel Hankey's remarks on the subject of special mechanical devices for taking trenches. It is extraordinary that the Army in the field and the War Office should have allowed nearly three months of warfare to progress without addressing their minds to its special problems.
The present war has revolutionised all military theories about the field of fire. The power of the rifle is so great that 100 yards is held sufficient to stop any rush, and in order to avoid the severity of the artillery fire, trenches are often dug on the reverse slope of positions, or a short distance in the rear of villages, woods, or other obstacles. The consequence is that war has become a short-range instead of a long-range war as was expected, and opposing trenches get ever closer together, for

mutual safety from each other's artillery fire.

The question to be solved is not, therefore, the long attack over a carefully prepared glacis of former times, but the actual getting across 100 or 200 yards of open space and wire entanglements. All this was apparent more than two months ago, but no steps have been taken and no preparations made.

It would be quite easy in a short time to fit up a number of steam tractors with small armoured shelters, in which men and machine guns could be placed, which would be bullet-proof. Used at night, they would not be affected by artillery fire to any extent. The caterpillar system would enable trenches to be crossed quite easily, and the weight of the machine would destroy all wire entanglements.

Forty or fifty of these engines, prepared secretly and brought into positions at nightfall, could advance quite certainly into the enemy's trenches, smashing away all the obstructions, and sweeping the trenches with their machine-gun fire, and with grenades thrown out of the top. They would then make so many *points d'appuis* for the British supporting infantry to rush forward and rally on them. They can then move forward to attack the second line of trenches.

The cost would be small. If the experiment did not answer, what harm would be done? An obvious measure of prudence would have been to have started something like this two months ago. It should certainly be done now.

The shield is another obvious experiment which should have been made on a considerable scale. What does it matter which is the best pattern? A large number should have been made of various patterns; some to carry, some to wear, some to wheel. If the mud now prevents the working of shields or traction engines, the first frost would render them fully effective. With a view to this I ordered a month ago twenty shields on wheels, to be made on the best design the Naval Air Service could devise. These will be ready shortly, and can, if need be, be used for experimental purposes.

A third device, which should be used systematically and on a large scale, is smoke artificially produced. It is possible to make small smoke barrels which, on being lifted, generate a great volume of dense black smoke, which could be turned off or on at will. There are other matters closely connected with this to which I have already drawn your attention, but which are of so secret a character that I do not put them down on paper.

One of the most serious dangers that we are exposed to is the possibility that the Germans are acting and preparing all these surprises, and that we may at any time find ourselves exposed to some entirely new form of attack. A committee of engineering officers and other experts ought to be sitting continually at the War Office to formulate schemes and examine suggestions, and I would repeat that it is not possible in most cases to have lengthy experiments beforehand.

If the devices are to be ready by the time they are required, it is indispensable that manufacture should proceed simultaneously with experiments. The worst that can happen is that a comparatively small sum of money is wasted.

      Yours etc.,
      Winston S. Churchill.

CHAPTER 2

Experiments

MARCH 1915 TO SEPTEMBER 1915

The first proposals put before the committee came from the Pedrail Transport Company of Fulham. In December 1914 this company had written to Mr. Churchill drawing his attention to a one-ton Pedrail machine which it had already constructed. This machine Mr. Churchill inspected on the Horse Guards Parade on February 16th, 1915.

On March 28th he confirmed his instructions to construct twelve Pedrails and six 16-foot Bigwheel Landships. The Pedrails were to be designed by Colonel R. E. Crompton and his assistant Mr. Legros, working from the design of the Fulham company's machine. The constructional work was placed in the hands of the Metropolitan Carriage Wagon and Finance Company Ltd.

Colonel Crompton was the well-known civil and mechanical engineer. He had seen service in the Crimean War, and during the South African War was engaged on mechanical traction problems.

At the time of the formation of the committee he was acting as consulting engineer to the Road Board, which, at the request of the Admiralty, very kindly released him in order that he might devote his attention to the landship problem.[1]

The Bigwheel Landships were to be designed by Mr. William Tritton of Messrs. Foster & Company of Lincoln, to whom the order was given on March 15th.

At this period I was acting as a Second Assistant to Major Hetherington. He asked me to come to the Admiralty to see Mr. d'Eyncourt, who wanted someone to drive the business forward as secretary. Mr. d'Eyncourt said he would be pleased with the arrangement. Hitherto,

---
1. His appointment terminated on August 31st, 1915.

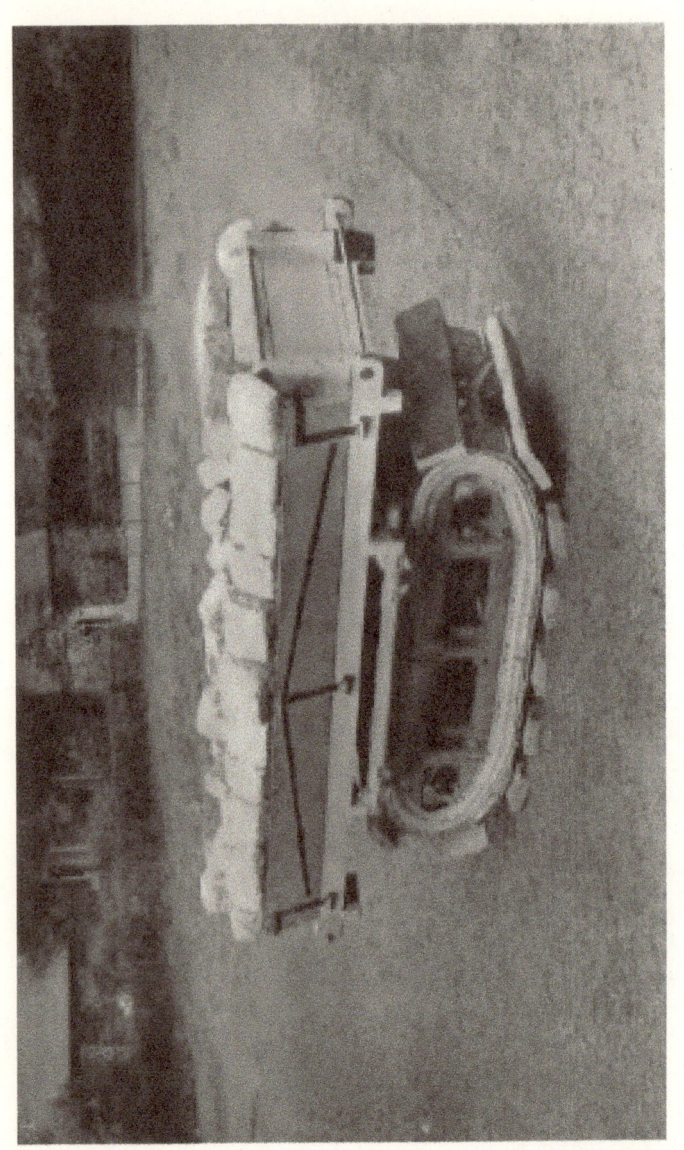

One-ton Pedrail machine as shown to Mr. Churchill in 1915

BIG WHEELER (MOCK UP)

Mr. P. Dale Bussell, of the Admiralty contract department, had been acting as secretary and had been responsible for orders and accounts. He now became a member of the committee, and later joined me at the Ministry of Munitions as my deputy. He rendered invaluable service from the start, until he was called to the Air Board, by very special request of the minister, to run their contracts in February 1917.

When I took over the duties of secretary of the landship committee in April 1916, Mr. d'Eyncourt was directing the affairs, assisted by Major Hetherington, who carried out his instructions, with Colonel Crompton as engineer.

On March 20th Colonel Crompton, Major Hetherington and I started for Neuve Chapelle to study and measure the captured German trenches. On arrival we discovered that Neuve Chapelle had been lost, and we had to return to England.

The question of armour was of great importance. Lieut. Kenneth Symes, who had been in charge of the building of the bodies for the armoured cars, was in charge, and remained responsible for all armour-plate until October 1917.

The original protection used by the armoured cars under Lieut.-Commander Samson consisted of two quarter-inch steel plates with an inch board between them. Even the German "S" bullet penetrated this as if it were butter.

Under the care of Lieut. Symes, assisted later by Lieut. W. E. Rendle, the light armour-plate made great strides, with the result that, in the end, the tanks hardly worried even about the German armour piercing bullet.

We tested a large number of patent materials. We never argued. Lieut. Symes's test was always sufficient—"just round the corner with a German rifle!"

Messrs. Beardmore gave most valuable help, making numerous test plates and experiments under the manager of their armour-plate department, Mr. T. M. Service. It is impossible to speak too highly of their help.

In June Mr. Churchill was leaving the Admiralty. Progress up till then had not been sufficiently marked. What had been done was explained in a report which we were then asked to make.

Admiralty Landships

These landships were at first designed to transport a trench-taking storming party of fifty men with machine-guns and ammunition; the men standing in two ranks at each side, and protected by side armour of 8 mm. thickness and roof armour of 6 mm. These vehicles were 40 feet long by 18 feet wide. The whole of the gear, engine and caterpillar supports were in the centre, allowing the side platforms on which the men stand to be only 18 inches above the ground level.

After a visit to France it was found that these long ships could not be easily steered round the narrow points and sharp road bends found in the villages close behind the firing line. The design was altered as follows:—

The vehicle was cut into two halves at mid length and articulated together by a special form of joint; on account of the increase of thickness of armour, found necessary by the increased penetrating power of the German bullet, from 8 mm. to 12 mm., the height of the side armour was reduced to 4 feet 6 inches. This compelled that the troops carried should be seated. The improved arrangement reduces the target presented to the artillery fire and has some other advantages.

In the new design the side armour is hinged at the top and is arranged so that it can be swung outwards in separate panels or as a whole; when in this position an armoured skirt falls down to a low level to give protection to the man working at the removal of wire entanglements. Both the side and the front armour is loopholed for rifle and machine-gun fire.

At first it was intended to adopt the only available form of caterpillar called the "Pedrail," but on subsequent examination it was found that this system had not been well developed for the larger size demanded for these landships, and a well-known and tried form of caterpillar, called the "Creeping Grip" system, has been ordered from America, and the first of these are expected to arrive shortly.

Preliminary experiments made on the marshes near Greenhithe have shown that the "Creeping Grip" type of caterpillar is well suited for this work, and sufficiently substantial in construction to stand all the hard usage it is likely to receive other than absolute destruction by shell fire.

The driving power for twelve of the landships is well in hand. In each case it consists of a pair of Rolls Royce engines, which for their reliability and silence in working appear to be without doubt the best existing type of engine, and it would have been impossible to develop any new type of engine of equal merit within a reasonable time.

As it appeared very important that the landships should be double-enders, that is to say, capable of working with equal facility in both directions, the whole are fitted with special type of reversing gear.

The engines are well in hand, and a supply for four ships have been inspected and are already on point of delivery to the Metropolitan Carriage Wagon and Finance Company, Ltd., who are carrying out the contract for the constructional work.

The work of the design and the preparation of the working drawings is now practically complete, with the exception of a few details connected with steering. The design of details relating to the armament, and to certain adjuncts, those which it is believed must be attached to the landships to render them efficient in every way, are yet under consideration. Among them are the following:—

1. The best appliances to adopt for the removal of such obstructions as wire-entanglements.

2. The form of mechanically worked trench-cutting device to be attached to the front of the Landships to enable them to dig themselves in as rapidly and as silently as possible; and

3. The class of armament required for grenade throwing or for projecting high explosives from points near the enemy's trenches, or the class of gun or howitzers, to be mounted on them in case they are required to carry armament fitted for more distant bombardment.

At a meeting under the presidency of Mr. Churchill on June 8th we explained that the "Creeping Grip" tractors mentioned in the report had been ordered from Chicago and inspected by our engineer out there, Lieut. Field, in order that we might carry out necessary experiments with a lighter track than the Pedrail.

The necessity of obtaining information from G.H.Q. about probable field obstructions and the difficulties to be overcome by landships was again discussed, and we agreed to approach, first, the director of fortifications and works at the War Office, General Scott Moncrieff. At this meeting, also, we decided to cancel Mr. Tritton's 16-feet wheel landship, for military reasons. It would have offered too large a target to the enemy.

On June 16th Mr. d'Eyncourt asked me to reorganise the committee on business lines. This was done and approved by Mr. d'Eyncourt.

Mr. A. J. Balfour now became First Lord of the Admiralty, and there was some doubt whether he would continue the work, which would become purely military. Our doubts were soon set at rest.

On the 22nd of June Mr. Balfour approved that the experimental work should proceed, as it was the policy of the late board and could not be abandoned without much loss. He also preferred not to take credit for any special schemes of the late First Lord, and requested Mr. Churchill to continue to interest himself in the Committee.

At this period no Government Department would provide any office accommodation for us, so on June 21st, 1915, I took an office at my own expense at 83, Pall Mall, and installed in it my entire organisation, which consisted of myself and Mr. Percy Anderson, at that time a petty officer in the Armoured Car Division.

A controversy raged on this subject for six months between the Admiralty, the Ministry of Munitions and the Office of Works.

The Admiralty referred to it as a troublesome case, and informed the Office of Works that a temporary Lieut. Albert G. Stern, R.N.V.R., had straightway proceeded to take an office for himself at 83, Pall Mall, and apparently did not understand the subtleties of the procedure in the civil service.

On June 23rd the landship committee was in possession of:—

(1) One Killen-Strait tractor at the armoured car headquarters at the Clement Talbot Works, Barlby Road;

(2) Two Giant Creeper Grip Tractors at the works of Messrs. McEwan, Pratt & Company, Ltd., Burton-on-Trent, which were to be coupled together for experiments;

(3) Two Diplock experimental one-ton wagons;

(4) An experimental ground at Burton-on-Trent.

At Burton-on-Trent we had Lieut. W. G. Wilson. He was the well-known engineer of the firm of Wilson-Pilcher (a pioneer firm in motorcar construction), and later of Messrs. Armstrong, Whitworth & Company. As a lieutenant in the Armoured Car Squadron he had helped in the construction of these cars, both the light car and the heavy car, which was armed with the 3-pounder naval gun. He was now supervising the experiments at Burton-on-Trent.

The committee was also engaged in the construction of two land battleships, one on Diplock Pedrails and the other on the special Creeper Grip tracks made in America.

Successful experiments had been carried out with the Killen-Strait tractor for cutting barbed wire, and experimental shields were being fitted on the 1½ ton Pedrail machines as a protection for advancing infantry. General Scott Moncrieff wrote to the committee:—

We think there should be at the bows of the landship on either side one 2-pounder pom-pom to deal with machine-gun emplacements.

That armament should be supplemented by two machine-guns placed further back, somewhat on the lines of the broadside fire of ships.

Loopholes for musketry fire would be required everywhere of course.

The above represents the view of the general staff here, but it may be modified in detail. However, if the details of the structures are arranged on this basis, it may be, I dare say, possible to introduce slight modifications later.

The 2-pounder pom-pom referred to above was made by Vickers Maxims and was an anti-aircraft gun. The 8-pounder naval gun had, up to now, not been found suitable for the destruction of houses and small forts, but orders had been given for high explosive shells for it, as the committee considered it would be the more suitable gun for its purposes.

A mortar, already used in France, was now at Barlby Road. It was thought that it might be suitable for destroying buildings and forts. This mortar threw a 50-pound shell 750 yards and a 25-pound shell 2000 yards. The shell was loaded with high explosive and had been found—especially the 50-pounder—very destructive indeed. The committee, however, were of opinion that the slowness of fire and the fact that the mortar was stationary would make it unsuitable for the landships.

At this stage it was considered advisable to request the War Office to arrange for their representative to attend our committee meetings in order to help, with their experience and criticism, the committee's plans for armour and armament, and the erection of obstacles at the experimental ground at Burton-on-Trent.

Mr. Balfour had asked Mr. Churchill to continue to act as chairman of the committee. At the same time the Admiralty handed over the control to the War Office, who appointed General Scott Moncrieff chairman.

I now had three chairmen, Mr. Churchill, General Scott Moncrieff, and Mr. d'Eyncourt. General Scott Moncrieff helped wherever he could and never interfered with, or obstructed, the pioneers who were struggling with this problem of constructing an armoured car that could cross all sorts of country, German trenches and perpendicular parapets, a problem which was said to be insoluble by nearly every technical expert on traction, and which the pioneers very nearly gave up as hopeless.

On the 29th of June the War Office and the Admiralty at last joined forces, although this had been opposed by Mr. Churchill since the start.

There was little to show at this period, and in order to raise enthusiasm amongst the army section I arranged for a display of different experimental apparatus at the armoured car headquarters on June 30th.

The *pièce de resistance* was wire-cutting by the Killen-Strait tractor with torpedo wire-cutters. Amongst those present were Mr. Lloyd George, Mr. Winston Churchill, Mr, d'Eyncourt, Sir Frederick Black, Major-General Scott Moncrieff, Colonel W. D. Bird, Colonel H. C. L. Holden and Brigadier-General Jackson.

Mr. Stokes of Ransomes & Rapier had shown me the Stokes gun on Clapham Common, which was being used as an experimental ground by General Louis Jackson, head of trench warfare at the Ministry of Munitions, and I had attended several private trials and was much impressed by the gun. I knew that Mr. Lloyd George, now minister of munitions, was coming out to see our trials, and so arranged with Mr. Stokes to have his gun at our old headquarters at Wormwood Scrubbs ready to fire for his benefit. Mr. Stokes had already convinced me of the impossibility of getting the War Office to take up the gun.

Mr. Lloyd George saw it fired with at least three shells in the air at the same time and was much impressed. He said to Mr. Stokes, "How long would it take to manufacture 1000 of these guns?"

Mr. Stokes replied, "It depends whether you or I try to make them." Anyhow, Mr. Lloyd George gave an order for 1000 of these guns, which have been one of the features of the war. That is one more debt of gratitude which the nation owes to Mr. Lloyd George.

On the 2nd of July, Squadron 20 of the Royal Naval Armoured Car Division, later to become famous as the "wet nurse" of tanks, was placed, for this work, under the direction of Mr. d'Eyncourt. The officers of the squadron were: Squadron-Commander Hetherington, Lieut.-Commander R. W. McGrath, Lieut. W. G. Wilson, Lieut. K. Symes and Lieut. A. G. Stern. Others were now added to it. The squadron was at Burton-on-Trent, and Squadron-Commander Hetherington remained in command. He had Lieut.-Commander Fairer-Smith, Lieut. W. P. Wilson and Lieut. W. G. Wilson as section leaders and three sub-lieutenants to make the full complement allowed for the Landship Squadron. I was to continue to act as secretary and attend to the execution of the instructions of the committee, and Lieut. Symes was attached for the special work on armour plate. Lieut.-Commander McGrath was shortly afterwards appointed second-in-command, instead of Lieut-Commander Fairer-Smith, and when Major Hetherington rejoined the Air Service early in 1916, he took the command and held it until the end of the war. Squadron 20, which at this date in 1915 numbered less than fifty, was by the end of the war at least twelve times as large.

This force, at a time when the army could spare no officers and men, tested and shipped all the tanks, carried out all experiments, including the first great experiment at Hatfield Park, and rendered most valuable service in numerous ways. This it continued to do until the end of the war.

In July, the Metropolitan Carriage, Wagon and Finance Company asked to be released from their contract to build the Pedrail machine, and the contract was transferred to Messrs. Foster & Company at Lincoln.

Colonel Crompton had put an enormous amount of work into his Pedrail design, but it had now to be abandoned owing to its great weight. We were faced with the fact that our original designs were both failures, and we had to set to work again.

The Pedrail material and the chains which had been specially manufactured were subsequently sold to the Trench Warfare Department, who had in view the building of a machine for carrying "*Flammenwerfer*" up to the enemy lines. The construction of this machine was carried out by a firm at Bath, the engines being supplied by the Astor Engineering Company. As certain mechanical defects developed, I believe that nothing further was done in the matter.

On July 29th Mr. Tritton was instructed to build a machine incorporating an armoured body, the 9-foot Bullock extended track and the 105 h.p. Daimler engine and transmission.

The machine was constructed as far as possible out of the parts at Mr. Tritton's disposal. The two 9-foot bullock tracks, the 105 h.p.

Daimler engine, the worm case, and the gear box were the same that had already been used for the heavy howitzer tractors. The body was made of boiler plate, so as to get the weight correct, but the turret, though of the correct weight, was a dummy, and would not revolve. The total gross weight of the machine was about eighteen tons, and the height to the top of the turret from the ground 10 feet 2 inches. The length of the tracks used on this machine were such that its power of crossing a sheer-sided trench was limited to a trench four feet wide.

Its speed was three-quarters of a mile an hour trench-crossing and two miles an hour when not negotiating severe obstacles. The steering was by a hand wheel operating two 4 foot 6 inch steering-wheels carried on a tail projecting six feet behind the body of the vehicle. The armament consisted of one 2-pounder automatic gun, with about 800 rounds of ammunition, one .303 Maxim, and several Lewis or Hotchkiss automatics, to be fired through the port-holes.

Mr. Tritton had all along been most anxious to help. He had spent his time and money in every way, and had already built for Admiral Bacon the 100 h.p. tractors for pulling the 15-inch howitzers and a trench-crosser. Besides the 16-foot Wheeler design, he had put forward an electrical Wheeler machine operated by cables, which had also been turned down.

Lieut. W. G. Wilson continued to help him and to act as Inspector for the committee.

The machine was first tried on September 6th in Messrs. Foster's yard. Certain mechanical defects showed themselves, but, on the 19th of September, the first trial over trenches took place on a hill outside Lincoln before Mr. d'Eyncourt, Mr. E. Moir, who was head of the recently formed inventions department of the Ministry of Munitions, Colonel E. D. Swinton, deputy secretary to the committee of imperial defence, Major Hetherington and myself.

This trial gave great satisfaction.

Already, in August, before this trial was held, Mr. Tritton and Lieut. Wilson had started to draw out a machine on the same lines but of stronger material and better design. On August 26th Mr. Tritton, Lieut. Wilson and I viewed the full-sized wooden model of this machine. It was known as the "Tritton" machine and later as "Little Willie." On the same day, at a meeting at the White Hart Hotel, Lincoln, we discussed fresh requirements which we had just received from the War Office. They asked that the machine should be able to cross a

KILLEN-STRAIT MACHINE

trench 5 feet wide with a parapet 4 feet 6 inches high. Lieut. Wilson and Mr. Tritton thereupon started work on a type designed to do this. It would, they told me, require a 60-foot wheel.

The contour of this sized wheel became more or less the shape of the underside of the new machine, which was called first the "Wilson" machine, then "Big Willie" and finally "Mother."

83, Pall Mall, S.W.
September 3rd, 1916.

Dear Tritton,

This is a private communication to you: I have also written to Wilson. It is of the utmost importance that we get on with the type known as the "Wilson." If it is a question of draughtsmen or designers we cannot allow any lack of assistance to stand in the way of our immediate progress. Please let me have a confidential note from you as to how we can speed this matter up.

'Tritton' Model

With reference to the gun in the turret: the ship will probably want to use its gun at an angle of 45 degrees pointing forward, upward or downward when crossing irregular ground and not on the flat, therefore it is necessary to be able to shift the gun forward as far as possible. It does not appear to me to be a difficult thing to run this on rails from the centre, where it is now, right forward.

I am also informed that it is not so necessary to protect the men from shrapnel, therefore a shield covering the front and the sides and the top partially seems to me to be sufficient.

I shall be very pleased to hear from you on the matter.

Yours faithfully,

Albert G. Stern.

This machine, to all intents and purposes, was, and remains, the heavy tank of today, Mark 5. But there were many difficulties to overcome. Endless experiments had to be made to produce a satisfactory track.

Now that we had progressed so far, after six months of secret experiments, excitement was intense to achieve a successful landship. Mr. d'Eyncourt, Mr. Tritton, Major Hetherington, Lieut. Wilson, Lieut. Symes, Mr. Anderson and I were all straining every nerve to succeed. It was the track that eluded us. Mr. d'Eyncourt turned down a proposed track of Balata belting, and once more our hopes sank. Then, on

September 22nd, I received the following telegram from Lincoln.

Stern, Room 59; 83, Pall Mall.

Balata died on test bench yesterday morning. New arrival by Tritton out of pressed plate. Light in weight but very strong. All doing well, thank you.

Proud Parents.

This was the birth of the tank.

CHAPTER 3

Trials—and Trials
September 1915 to February 1916

I have already spoken of the impossibility of finding any government department which would give us accommodation. That was only one of our many difficulties. We encountered opposition from all quarters. Manufacturers did not like our type of work. It was all experimental and meant continual cancelling of orders. Then, in July, the Ministry of Munitions took over all inventions in connection with land warfare, and the Admiralty, quite rightly, was unwilling to provide the men for these experiments. This meant the loss of Squadron 20, and without Squadron 20 all our experiments and transport would have stopped.

In August the whole of the Armoured Car Division was disbanded, and on the 18th of that month the following order was issued:—

Will you please inform the president of landships committee that, in accordance with instructions received from the Admiralty, No. 20 Squadron is being recalled to headquarters with a view to transfer of officers and men to the army or air service.

At the same time Squadron 20 got its sailing orders.

You are to return with your squadron to headquarters at 9.30 a.m. on Friday next, 20th inst., to bring with you all gear and stores which have not been purchased by the landships committee. All officers and men are to be present.

This disbandment was stopped by the personal intervention of Mr. d'Eyncourt. It was one of the many occasions on which he saved the landships (and future tanks) from extinction.

I also made a personal request to the Minister of Munitions, and was told by him that the Admiralty informed him that the order was to be disregarded.

On August 19th Mr. d'Eyncourt wrote to the Second Sea Lord as follows:—

> Officers and men of Squadron 20 (Armoured Car Division) who have been detailed for duty under the landships committee, have been offered the choice of joining the army and have refused.
>
> In view of this it appears that there can be no objection to the request of the Minister of Munitions being granted; otherwise the work on the landships will come to an immediate standstill.

The Second Sea Lord replied on the same date:—

> I have approved as above and all tools, etc., required by the Committee should be left with them until other arrangements can be made.

Towards the end of September another attempt was made to get rid of Squadron 20, and in the end an arrangement was made which was set out by the Admiralty in a document on October 1st. The landships committee of the Admiralty was to take its instructions as to type, armament and protection from the War Office, was to carry on the designing and constructing as far as was desirable, was then to hand over the work to the Ministry of Munitions, and was, for this purpose, to make use of the personnel, transport and supplies which were at its disposal— that is to say. No. 20 Squadron of the R.N.A.S.

On the 20th of October the First Lord, Mr. Balfour, approved the minute.

Mr. Macnamara then suggested, for secrecy's sake, to change the title of the landships committee. mr. d'eyncourt agreed that it was very desirable to retain secrecy by all means, and proposed to refer to the vessel as a "water carrier." In Government offices, committees and departments are always known by their initials. For this reason I, as secretary, considered the proposed title totally unsuitable. in our search for a synonymous term, we changed the word "water carrier" to "tank," and became the "tank supply," or "T.S." committee. This is how these weapons came to be called "tanks," and the name has now been adopted by all countries in the world.

Although the Ministry of Munitions had now taken over from the Admiralty all inventions which were to be used solely for military purposes, it was arranged that the landships committee should continue as hitherto under the chairmanship of Mr. d'Eyncourt, and that progress should be reported periodically to Mr. Moir of the munitions inventions department.

Mr. Moir was at all times most helpful. He refrained from interfering in a development which was now in full swing. Under this new arrangement the commanding officer of Squadron 20 took command of the Ministry of Munitions experimental ground at Wembley and gave all the necessary help for all experiments carried out there. This included all "*Flammenwerfer*" or flame projectors, and many other experiments of the trench warfare department. This was a very satisfactory arrangement. We were looking for an experimental ground, when I found General Louis Jackson, director of trench warfare, with a ground and no men. We had Squadron 20 but no ground. So we united, with a large saving of money to the nation; but there were several hurdles of red tape to negotiate before the thing was done.

It was about this time—July 1915—when we were struggling with the mechanical problem and fighting to be allowed to exist, that Colonel Swinton, by chance, discovered us. Years ago—I believe in 1908, or perhaps a little earlier—he or his friend Captain Tulloch, who was an expert on ballistics and high explosives, put before the military authorities ideas of mechanical warfare which, I am told, are still filed at the War Office. At the beginning of the war he again put forward his ideas, this time at G.H.Q., France. They were sent in to the War Office, and there the experts decided that from a technical and manufacturing point of view the weapons which he proposed were impossible. They would take such a long time to design that the war would be over before any could be manufactured.

He was now to play an important part in getting the military authorities to take up the idea of tanks. With his keen sense of humour, understanding of the value of propaganda, intimate knowledge of the War Office and all its mysterious ways, and with his exceptional position as deputy secretary to the committee of imperial defence, he was able to push forward our schemes and to cut short all sorts of red tape for us. It was largely owing to his efforts that the army took up tanks and developed the tactics rapidly enough to make it impossible for the German Army ever to catch us up.

He asked me to go and see him at the committee of imperial defence.

"Lieut. Stern," he said, "this is the most extraordinary thing that I have ever seen. The Director of Naval Construction appears to be making land battleships for the army who have never asked for them, and are doing nothing to help. You have nothing but naval ratings doing all your work. What on earth are you? Are you a mechanic or a chauffeur?"

"A banker," I replied.

"This," said he, "makes it still more mysterious."

It was Colonel Swinton who got the prime minister to call an inter-departmental committee on August 28th.

This committee went very thoroughly into the whole question of procedure in connection with future experiments with landships, and as a result many difficulties which we had hitherto experienced were swept away.

The following letter from Mr. d'Eyncourt explains itself:—

73, Cadogan Square, S.W.,
August 29th, 1915.

My dear Stern,

Many thanks for your letter and copy of agenda; the conference has distinctly cleared the air and put the whole thing on a sounder footing. I'm glad you had a good talk with Swinton; you seem to have arranged it very well, and I now hope we shall be able to go on steadily without more tiresome interruptions.

I hope Foster is getting on well and will be able to push on the new design also.

We must be very careful about secrecy, especially in conveying the things to Wembley, and in keeping everything quiet there; we should have a high fence put round, and all the men of Squadron 20 must be trustworthy and specially warned about talking.

Foster's people should also be warned; will you write him a letter impressing the importance of secrecy? I think I did speak to him about it, but another letter won't do any harm.

I shall be at Admiralty again on Wednesday. I am keeping your letter as a minute of the conference.

Very sincerely yours,
E. H. T. d'Eyncouet.

Secrecy for our new weapons was all-important. Everybody connected was sworn to secrecy. Anybody suspected of talking was threatened with internment under D.O.R.A. Ladies sometimes were found to have heard something about us, and had to be visited and told it would cost thousands of lives if the secret reached the enemy. Flying men had to be forbidden to fly over the tank grounds. It was easy to stop all talk amongst those who knew of the start of our enterprise by informing them that our efforts had entirely failed and that we had lost our jobs, which they were only too ready to believe.

On September 20th I arranged for Lieut. Symes to bring by motor lorry the complete full-sized model of "Mother" to the Trench Warfare Experimental Ground at Wembley, the headquarters of Squadron 20, and on September 22nd orders were given to Mr. Tritton, of Foster's, to continue working on the design of "Mother" with all dispatch, but not to stop rebuilding the first type called "Little Willie."

The director of staff duties at the War Office, Colonel Bird, now informed the commander-in-chief of the British Army in the field that a wooden model of the proposed landship would be ready for inspection on Wednesday, September 29th, and asked him to send representatives. This meeting took place, and there were present Mr. d'Eyncourt, Major Hetherington, Lieut. W. G. Wilson, Lieut. Stern and Lieut. K. P. Symes of the committee; Major Segrave, Colonel Bird and Colonel Holden of the War Office; Colonel Harvey, Major Dryer and Major Guest from G.H.Q., France; Colonel Goold Adams and Captain Hopwood from the Ordnance Board; Brig.-General Jackson, Mr. Moir, Captain Tulloch, Captain Acland, Mr. Tritton and Colonel Swinton.

All expressed their satisfaction with the display, but there were others not so well pleased. The minutes of the meeting were duly circulated to all concerned, and General Von Donop strongly expressed his disapproval of the procedure which had been adopted. He viewed with dismay the fact that the War Office, the committee of imperial defence and the Admiralty were all mixed up in deciding this question. He was also somewhat annoyed that he should have been asked to provide guns and ammunition when he had not been consulted as to their pattern.

The officials concerned generally were getting rather sceptical about the progress of our committee, and on September 30th I gave Mr. Tritton instructions to carry on with all possible speed the construction of "Mother."

In the design of this machine, with the track running over the top, the difficulty was the position of the guns, and Mr. d'Eyncourt gave me instructions, which I forwarded to Mr. Tritton on October

5th, giving the size of the opening of the "sponsons,"[1] which would carry one gun on each side of the proposed ship, and the weight of the gun-carriage, of the base-plate, and of the holding-down ring and shield. These two sponsons, which carried two guns and shields, were to weigh in all about three tons.

At this time it had not been finally decided what gun we should have, although the sponsons were designed to carry the 6-pounder. A 2.95 inch mountain gun was borrowed from Woolwich and proved satisfactory, but it was impossible to get, so in the end the 6-pounder naval gun was adopted. Admiral Singer having told us that he was ready to release a number of these guns. Thus another vital difficulty was overcome by the Admiralty.

At the end of October Foster's workmen were leaving the firm. It had frequently applied to the Government for war badges for them, but had never been able to get them, and now, owing to the great secrecy of the work on which the men were employed, their comrades thought that they were not doing war work. I found it impossible to obtain the badges, until at last I went to the offices of the War badge department in Abingdon Street and threatened to take them by force with the aid of Squadron 20. Thereupon a sack of badges was delivered. On October 29th Mr. Tritton wrote to me:—

> I am very grateful to you for the trouble you have taken in the matter, and I feel I must congratulate you on the promptness with which you have overcome the entanglement of red tape which is apparent in Abingdon Street.

On November 22nd the first of the tracks for "Little Willie" was completed and ran in the shop on its own axle. On account of the shortage of material, the second track could not be completed till later. "Mother's" hull was expected to be in the erecting shop the next day, and the engine and gearing were well on the way to completion. Owing to a shortage of links, no approximate promises for completion were given, but a promise was expected at the end of the week.

On December 3rd the first trials of "Mother" took place at Lincoln, and were very successful indeed. It was hoped to have a machine on the road by December 20th, and to bring it to London for the trials a fortnight later.

---

1. A sponson is a structure projecting beyond the side of a tank or ship in which a gun is placed. This projection is necessary to enable the gun to be fired clear of the side, directly ahead or astern.

Mr. d'Eyncourt also wished to have a 6-pounder gun fired from the finished tank in order to see what effect it would have on the sponson, the frame, and the crew. "Mother" was therefore moved to a lonely field within a mile of Lincoln Cathedral. The night before Major Hetherington and I motored up with the ammunition, which was solid, armour-piercing, 6-pounder shell. Early next morning, everything being ready, Major Hetherington fired the first shot. There was a misfire, and while we were examining the breech the gun went off itself. No one knew where the shell had gone. We feared the worst. Lincoln Cathedral was in danger! But after two hours spent with a spade the shell was found buried in the earth, to the great relief of us all.

It was quite clear that the experimental ground at Wembley was not sufficiently private for these very secret trials which we intended now to carry out, so Mr. d'Eyncourt and I started very early one morning looking for a suitable place north of London, and finally saw Lord Salisbury's agent at Hatfield, Mr. McCowan. We selected a certain part of the park for the trial, and Mr. McCowan gave us every facility. Lord Salisbury afterwards gave the necessary permission.

For the lighter armament of the tanks we tried every machine-gun to see which was the most suitable—the Lewis gun, the Hotchkiss gun, the Vickers and the Madsen, which were all lent to us by Colonel Brown of the Enfield Lock Small Arms Factory. The Hotchkiss was eventually selected.

Colonel Swinton was convinced that there was a great future for mechanical warfare, and in order that everything should be ready, if the experiments were successful, he had arranged for an inter-departmental conference of the committee of imperial defence to be held on December 24th.

This committee was held and recommended that, if and when the army council (after inspection of the final experiments on the land cruisers) decided that they would be useful to the army, the provision of these machines should be entrusted to a small executive supply committee—which for secrecy's sake should be called the "tank supply committee"—to come into existence as soon as the decision of the army council was made; that in order to carry out its work with the maximum of dispatch and the minimum of reference, it should have full power to place orders and correspond direct with any government departments concerned; that it should be composed of those who had hitherto carried on the work, and that the War Office should

CREEPING GRIP MACHINES COUPLED TOGETHER

CREEPING GRIP MACHINE

KILLEN-STRAIT MACHINE FITTED WITH WIRE CUTTING APPARATUS

KILLEN-STRAIT MACHINE, FIRST FITTED WITH WIRE-CUTTING APPARATUS. MR. LLOYD GEORGE AND MR. WINSTON CHURCHILL.

take the necessary steps to raise a body of suitable men to man the land cruisers.

The committee made its recommendations in great detail, of which the above is only a *résumé*, and these recommendations were submitted to the Admiralty, which agreed to give every facility: (1) By lending Mr. d'Eyncourt, who had an intimate knowledge of all that had been done, and who expressed his willingness to give all the assistance he could; (2) by supplying 6-pounder guns; (3) by transferring the officers and men of Squadron 20 to the army.

The first tank, "Mother," was finished on January 26th, 1916, and sent by train to Hatfield Station, where it was unloaded in the middle of the night and driven up to the special ground in Hatfield Park. A detachment of Squadron 20, under the Command of Major Hetherington, had previously been sent to Hatfield.

Large numbers of the 3rd (Mid Herts) Battalion Herts Volunteer Regiment and a company of engineers, lent by the War Office, helped to dig the necessary trenches for the trials, the first of which took place on the 20th of January, 1916.

The following were present—

E. H. Tennyson d'Eyncourt, Esq., C.B.
General Sir G. K. Scott Moncrieff, K.C.B., C.I.E.
Rear-Admiral F. C. T. Tudor, C.B.
Rear-Admiral Morgan Singer, C. B.
Brigadier-General L. C. Jackson, C.M.G.
Brigadier-General H. C. Nanton, C.B.
Brigadier-General Hill, C.B.
Colonel R. E. Crompton.
Colonel W. D. Bird, C.B., D.S.O.
Colonel Harrison, R.A.
Colonel F. B. Steel.
Lieut.-Colonel M. Hankey, C.B.
Lieut.-Colnel E. D. Swinton, D.S.O.
Lieut.-Colonel W. Dally Jones.
Lieut.-Colonel Byrne.
Lieut.-Colonel G. H. S. Browne.
Lieut.-Colonel Byron, R.E.
Lieut.-Colonel Matheson, R.E.
Commodore Murray F. Sueter, C.B.
Major Lindsay.
Major S. T. Cargill, R.E.

Captain T. G. Tulloch.
Captain C. A. Bird, R.A.
Flight-Commander H. A. Nicholl.
Lieut.-Commander P. B. Barry.
Dr. C. Addison.
F. Kellaway, Esq., M.P.
P. Dale Bussell, Esq.
Sir Charles Parsons.
Mr. McCowan (agent to Lord Salisbury).
Mr. W. A. Tritton.
Mr. Yeatman.
Mr. Broughton
Mr. Starkey
Mr. Sykes of Messrs. Foster and Company.
Major Hetherington.
Lieut. W. G. Wilson.
Lieut.-Commander R. W. McGrath.
Lieut. K. Symes.
Lieut. W. P. Wilson.
Lieut. G. K. Field.
Lieut. A. G. Stern.

The Programme of the First Trial was as follows:—

TANK TRIAL

DESCRIPTION OF A "TANK"

This machine has been designed, under the direction of Mr. E. H. T. d'Eyncourt, by Mr. W. A. Tritton (of Messrs. Foster of Lincoln) and Lieut. W. G. Wilson, R.N.A.S., and has been constructed by Messrs. Foster of Lincoln. The conditions laid down as to the obstacle to be surmounted were that the machine should be able to climb a parapet 4 feet six inches high and cross a gap 5 feet wide.

OVER-ALL DIMENSIONS

	Feet.	Inches.
Length	31	3
Width with sponsons	13	3
" without sponsons	8	3
Height	8	0

PROTECTION

The conning-tower is protected generally by 10 mm. thickness of nickel steel plate, with 12 mm. thickness in front of the drivers. The sides and back ends have 8 mm. thickness of nickel-steel plate. The top is covered by 6 mm. thickness of high tensile steel and the belly is covered with the same.

Weight

	Tons.	Cwts.
Hull	21	0
Sponsons and guns	3	10
Ammunition, 300 rounds for guns and 20,000 rounds for rifles (removable for transport purposes)	2	0
Crew (8 men)	0	10
Tail (for balance)	1	8
Total weight with armament, crew, petrol and ammunition	28	8
Horse Power of engines	105 h.p.	
Number of gears	4 forward, 2 reverse.	
Approximate speed of travel on gear	¾ mile, 1¼ miles, 2¼ miles, and 4 miles per hour.	

Armament

Two 6-pounder guns, and
Three Automatic rifles (1 Hotchkiss and 2 Madsen).

Rate of Fire

6-pounder, 15 to 20 rounds per minute.
Madsen gun, 300 rounds per minute.
Hotchkiss gun, 260 rounds per minute.

Notes as to Steel Plate

obtained from Experiments Made Nickel-Steel Plate

12 mm. thickness is proof against a concentrated fire of reversed Mauser bullets at 10 yards range, normal impact.

10 mm. thickness is proof against single shots of reversed Mauser bullets at 10 yards range normal impact.

8 mm. thickness is proof against Mauser bullets at 10 yards range, normal impact.

PLAN

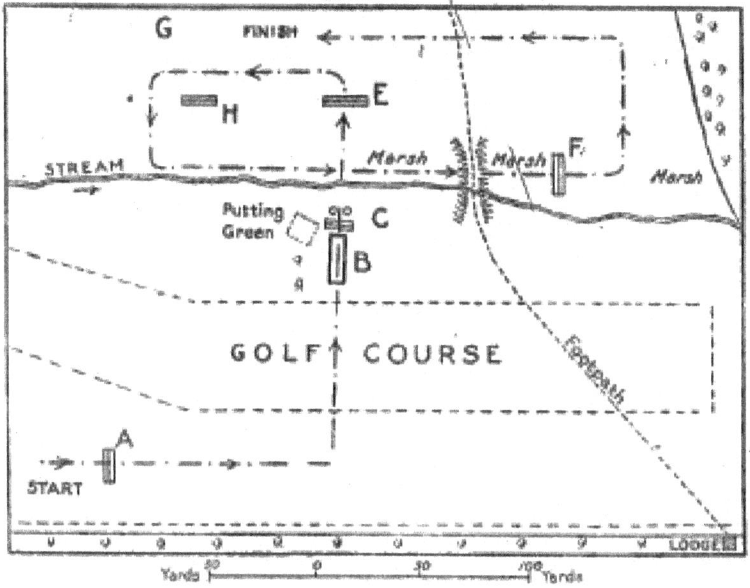

SECTIONS.

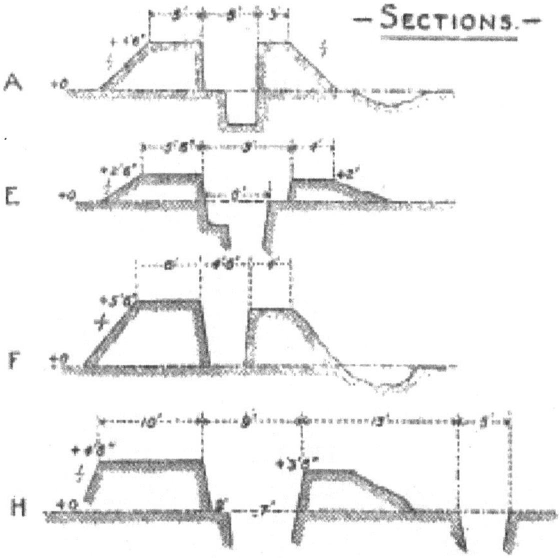

High Tensile Steel Plate

6 mm. thickness will give protection against bombs up to 1 lb. weight detonated not closer than 6 inches from the plate.

N.B.—It is proposed to cause the detonation of bombs away from the top of the Tank by an outer skin of expanded metal, which is not on the sample machine shown.

Programme of Trials

Reference to Sketch, Plan and Sections

The trial will be divided into three parts, 1., 2., and 3.

Part 1.—*Official Test*

1. The machine will start and cross (a) the obstacle specified, *i.e.* a parapet 4 feet 6 inches high and a gap 5 feet wide. This forms the test laid down.

Part 2.—*Test approximating to Active Service*

2, It will then proceed over the level at full speed for about 100 yards, and take its place in a prepared dugout shelter (b) from which it will traverse a course of obstacles approximating to those likely to be met with on service.

3. Climbing over the British defences (c) (reduced for its passage), it will—

4. Pass through the wire entanglements in front;

5. Cross two small shell craters, each 12 feet in diameter and 6 feet deep;

6. Traverse the soft, water-logged ground round the stream (d), climb the slope from the stream, pass through the German entanglement.

7. Climb the German defences (e).

8. Turn round on the flat and pass down the marshy bed of the stream *via* (d) and climb down the double breastwork at (f).

Part 3—*Extra Test if required*

9. The "tank" will then, if desired, cross the larger trench (h) and proceed for half a mile across the park to a piece of rotten ground seamed with oil trenches, going down a steep incline on the way.

January 27th. 1916.

The day after this first trial Mr. d'Eyncourt wrote to Lord Kitchener:—

January 30th, 1916.

My Lord,

As the head of the Admiralty committee entrusted with the design and manufacture of a trial machine to cross the enemy trenches, carrying guns large enough to destroy machine-guns, and to break through wire entanglements whilst giving protection to its own crew—conditions laid down by the War Office—I now have the honour to report that after much experimental work and trying several types, we have produced a machine complete with armament which amply fulfils all the requirements.

This machine has had a satisfactory preliminary trial at Hatfield and proved its capacity, and I trust your Lordship may be able to come there and see a further trial on Wednesday afternoon next, February 2nd, when you will have an opportunity of judging of its qualities yourself.

I have had some experience in the production of war material for H.M. Navy, and I am convinced that the machine you will see is capable of doing its work—work never before accomplished—and though as the first really practical one of its kind, it can no doubt be improved and considerably developed, yet as time is so important, and it will take three or four months to produce them in sufficient numbers, I venture to recommend that, to prevent delay, the necessary number be ordered immediately to this model without serious alteration. While these are being manufactured we could proceed with the design and production of more formidable machines of improved type, with such modifications as your Lordship might approve.

I have the honour to be, etc. etc.

E. H. T. d'Eyncourt.

Field Marshal,
The Right Hon. Earl Kitchener, War Office.

On February 2nd, 1916, the second trial of the machine took place, when the following were present:—

Field-Marshal the Right Hon. H. H. Earl Kitchener of Khartoum, K.G., G.C.B., O.M., G.C.S.I., G.C.M.G., G.C.I.E. (Secretary of State for War).

The Right Hon. A. J. Balfour, M.P.
The Right Hon. D. Lloyd George, M.P.
The Right Hon. R. McKenna, M.P.

E. H. Tennyson d'Eyncourt, C.B.
Vice-Admiral Sir Frederick T. Hamilton, K.C.B., C.V.O.
Sir W. Graham Greene, K.C.B.
Commodore D. M. de Bartolome, C.B.
The Right Hon. G. Lambert, M.P.
Brigadier-General C. E. Corkran, C.M.G.
Major-General Butler.
Major-General Sir S. B. Von Donop. K.C.B. (Master General of the Ordnance).
Major-General H. G. Smith. C.B.
Lieut.-General Sir John S. Cowans, K.C.B., M.V.O. (Quarter-Master General to the Forces).
Brigadier-General H. C. Nanton, C.B.
General Rudyear.
Lieut.-General Sir W. R. Robertson, K.C.B., K.C.V.O., D.S.O. (Chief of the Imperial General Staff).
Major-General R. D. Whigham, C.B., D.S.O. (Deputy-Chief of the Imperial General Staff).
Brigadier-General F. B. Maurice, C.B. (Director of Military Operations).
Bt. Lieut.-Colonel A. A. G. FitzGerald, C.M.G. (Private Secretary to Secretary of State for War).
The Hon. E. FitzGerald.
Colonel H. E. F. Goold Adams, C.M.G.
Lieut.-Colonel G. L. Wheeler.
Lieut.-Colond C. Evans, C.M.G.
Colonel A. Lee, M. P.
Major C. L. Storr.
Major Segrave.
Major H. O. Clogstoun, R.E.
Captain T. G. Tulloch.
Dudley Docker, Esq.
Major Greg.
J. Masterton Smith, Esq.
Lieut.-Colonel M. Hankey, C.B.
Lieut.-Colonel E. D. Swinton, D.S.O.
Mr. F. Skeens.
Major T. G. Hetherington.
Lieut. W. G. Wilson.
Lieut.-Commander R. W. McGrath

Lieut. K. Symes.
Lieut. W. P. Wilson.
Lieut. Donnelly.
Lieut. G. K. Field.
Mr. W. A. Tritton.
Lieut. A. G. Stern.

The Right Honourable H. H. Asquith, M.P., was unable to attend.

Colonel Hankey arranged for Mr. McKenna, Chancellor of the Exchequer, to travel down to the Hatfield trials in my car. I explained to him our ideas of mechanical warfare and its value in the saving of life and shells. After the trials Mr. McKenna said that it was the best investment he had yet seen, and that if the military approved, all the necessary money would be available.

Mr. Balfour, amongst others, took a ride in the tank, but was removed by his fellow ministers before the machine tried the widest of the trenches. This was a trench more than nine feet wide which Lord Kitchener wished to see it cross, but which it had never attempted before.

As Mr. Balfour was being removed feet first through the sponson door he was heard to remark that he was sure there must be some more artistic method of leaving a tank!

Sir William Robertson was well satisfied with the machine. He left the ground early, owing to pressing business, but before he went he told me that orders should be immediately given for the construction of these machines.

General Butler, who was at this time deputy chief of the general staff to Sir Douglas Haig at the front, asked how soon he could have some of them and what alterations could be made. I told him that no alterations could be made if he wished any machines that year, except to the loop-holes in the armour-plate.

On February 8th His Majesty the King visited Hatfield, when a special demonstration was arranged. He took a ride in the tank, and said afterwards that he thought such a weapon would be a great asset to the army possessing a large number.

A few days later Mr. d'Eyncourt wrote to Mr. Churchill, now a colonel at the front:—

Admiralty, S.W.
February 14th, 1916.

Dear Colonel Churchill,

It is with great pleasure that I am now able to report to you the success of the first landship (tanks we call them). The War Office have ordered one hundred to the pattern which underwent most successful trials recently. Sir D. Haig sent some of his staff from the front. Lord Kitchener and Robertson also came, and members of the Admiralty board. The machine was complete in almost every detail and fulfils all the requirements finally given me by War Office. The official tests of trenches, etc., were nothing to it, and finally we showed them how it could cross a 9 feet gap after climbing a 4 feet 6 inches high perpendicular parapet. Wire entanglements it goes through like a rhinoceros through a field of corn. It carries two 6-pounder guns in sponsons (a *naval* touch), and about 300 rounds; also smaller machine-guns, and is proof against machine-gun fire.

It can be conveyed by rail (the sponsons and guns take off, making it lighter) and be ready for action very quickly. The King came and saw it and was greatly struck by its performance, as was everyone else; in fact, they were all astonished. I wish you could have seen it, but hope you will be able to do so before long. It is capable of great development, but to get a sufficient number in time, I strongly urged ordering immediately a good many to the pattern which we know all about. As you are aware, it has taken much time and trouble to get the thing perfect, and a practical machine simple to make; we tried various types and did much experimental work.

I am sorry it has taken so long, but pioneer work always takes time, and no avoidable delay has taken place, though I begged them to order ten for training purposes two months ago. After losing the great advantage of your influence I had some difficulty in steering the scheme past the rocks of opposition and the more insidious shoals of apathy which are frequented by red herrings, which cross the main line of progress at frequent intervals.

The great thing now is to keep the whole matter secret and produce the machines all together as a complete surprise. I have already put the manufacture in hand, under the *ægis* of the minister of munitions, who is very keen; the Admiralty is also allowing me to continue to carry on with the same committee, but Stern is now chairman.

I enclose photo. In appearance, it looks rather like a great antediluvian monster, especially when it comes out of boggy ground, which it traverses easily. The wheels behind form a rudder for steering a course, and also ease the shock over banks, etc., but are not absolutely necessary, as it can steer and turn in its own length with the independent tracks.

In conclusion, allow me to offer you my congratulations on the success of your original project and wish you all good luck in your work at the front.

E. H. T. d'Eyncourt.

    Colonel W. S. Churchill,
    6th Royal Scots Fusiliers,
    B.E.F., France.

Colonel Churchill replied about a week later, saying how pleased he was and that he would like to see the machine. I wrote on March 3rd, saying we could show it him at Lincoln.

As a result of the trials, Rear-Admiral F. C. B. Tudor, the third sea lord, reported:—

I am convinced, if these tanks are to be built in any reasonable time, it can only be done by putting the matter in the hands of an independent executive committee with authority to order material and incur expenditure up to a fixed limit.

I saw the demonstration in Hatfield Park on 29th inst. and was much impressed; the tank carried out the official test with the greatest ease and also many other seemingly impossible tasks; in fact, it is probably the only solution of the stalemate of trench warfare.

To my mind the right course would be to put in hand fifty machines exactly similar to the experimental 'tank' immediately, spreading the orders for the various parts required over a large number of firms, as we have done in the case of submarine engines, and at the same time get out fresh designs for a much larger tank, perhaps three times the length, with 3 feet protection over forepart, the armament weights being not necessarily increased.

Such a machine would have no difficulty in crossing heavily-traversed trenches from any line of approach, without fear of capsizing; in fact, there is practically no limit to the development of this class of machine, except the cost.

This was brought before the Admiralty board, concurred in by Mr. Balfour and Admiral Jackson, and the congratulations of the board to the director of naval construction were stamped and signed by Sir William Graham Greene.

Finally, the army council wrote to the Admiralty on February 10th:—

Sir,

I am commanded by the Army Council to request that the Lords Commissioners of the Admiralty will convey the very warm thanks of the army council to Mr. E. H. T. D'Eyacourt, C.B., director of naval construction, and his committee, for their work in evolving a machine for the use of the army, and to Mr. W. A. Tritton and Lieut. W. G. Wilson, R.N.A.S., for their work in design and construction. I am to state that their efforts in this connection have been highly appreciated by the army council.

I am to add that the work of the officers and men of No. 20 Squadron, R.NA.S., in assisting by constructing trenches, etc., in the experimental work necessary to the production of the machine has been of valuable service.

     I am. Sir,
       Your obedient Servant,
          B. B. Cubitt.

The thanks of the War Office are thoroughly well-deserved, wrote Mr. Balfour.

CHAPTER 4

First Orders

FEBRUARY 1916 TO SEPTEMBER 1916

After the trial at Hatfield, Lord Kitchener asked me to go to the War Office as head of the department which the inter-departmental conference of the committee of imperial defence had recommended should be set up for the production of tanks. I saw Major Storr, Lord Kitchener's Secretary, and later had two interviews with Sir Charles Harris, the assistant financial secretary of the War Office.

In the meantime Mr. Lloyd George sent for me and asked me if I would go to the Ministry of Munitions. I told him that Lord Kitchener had already asked me to go to the War Office, and he replied that it was not a matter for the War Office but for the Ministry of Munitions. I said that I was willing to undertake the production of tanks in quantity within six months, but could only do so if given special powers. Mr. Lloyd George then asked me to write out a charter which, if approved, he would sign. With the aid of Colonel F. Browning of the Ministry of Munitions, and Mr. P. Dale Bussell, I drafted the following charter, which gave me exceptional powers. It was signed by Mr. Lloyd George on February 12th, 1916, subject to one paragraph being approved by Mr. Sam H. Lever, who was at that time assistant financial secretary.

> The minister of munitions has had under consideration the report of the inter-departmental committee on the question of tank supply.
> The minister considers that now the question of design has been settled and it has been decided to arrange for earliest possible supply of 100 tanks, the matter becomes one of supply which falls within the province of the Ministry of Munitions

53

PEDRAIL MACHINE BUILT AT BATH

Sir William Tritton

to arrange.

As, however, the Admiralty committee has carried out the whole of the experimental construction and is fully acquainted with the details necessary for the production of a large number of tanks, the minister agrees with the first lord that the committee should now become a committee attached to the department of minister of munitions and should carry on the work and arrange for the manufacture of the machines required as an executive body working directly under the minister of munitions.

The composition of the committee will be as shown on attached sheet, and the Admiralty has agreed to allow Admiralty members of the committee to undertake these duties. Mr. d'Eyncourt, director of naval construction, has consented to continue to superintend the technical and experimental work of the committee.

The minister accordingly authorises the committee to arrange manufacture of these machines, placing orders with contractors as necessary and corresponding direct any government department concerned, also to incur any necessary expenditure in connection with engagement and remuneration of inspecting or other staff, experimental work, travelling and other incidental expenses. The committee shall have the final decision in all matters connected with the manufacture and inspection of these machines, and shall have full power to depute to any one of their number any specific duties concerned with the above, and also to add to their number if necessary.

The Minister of munitions will grant all facilities required by the committee for supply of labour and material to the contractors for the tanks. All payments for this work shall be made solely on the certificate of the committee which shall be accepted as full and sufficient authority by all departments concerned, and an imprest[1] of £50,000 is to be at once placed at the committee's disposal for the experimental work.

D. L. G.
12/2/16.

---

1. I concur subject to the word 'imprest' being changed to 'authorisation.'

Sam H. Lever,
D. of F.
12/2/16.

The original constitution of the tank supply committee was as follows:—

Lieut. A. G. Stern, R.N.A.S., Director of Naval Construction's Committee (Chairman).

E. H. T. d'Eyncourt, Esq., C B., Director of Naval Construction.

Lieut.-Colonel E. D. Swinton, D.S.O., R.E., Assistant Secretary, Committee of Imperial Defence.

Major G. L. Wheeler, R.A. Director of Artillery's Branch, War Office.

Lieut. W. G. Wilson, R.N.A.S., Director of Naval Construction's Committee.

Lieut. K. Symes, R.N.A.S., Director of Naval Construction's Committee.

P. Dale Bussell, Esq., Director of Naval Construction's Committee, Contract Department, Admiralty.

A day or two later Lieut.-Colonel Byrne and Captain T. G. Tulloch of the Ministry of Munitions were added to the committee.

The committee formed, Mr. Lloyd George then told me to get rooms for it at the Ministry of Munitions. Secrecy was the essence of tanks. In all the business of the landship committee, and afterwards the tank supply department, everything possible was done—and successfully—to keep the tanks a secret. This resulted in many difficult situations. When I went across to the ministry to get rooms after my interview with Mr. Lloyd George, I was refused, and my whole department treated as a joke, owing to the fact that we were not allowed to explain what our business was. As a result I was forced to appropriate rooms, as the following minute from Sir Frederick Black (at that time director-general of munitions supply) to Sir Arthur Lee, the parliamentary military secretary, will show.

> I understand that rooms have been appropriated in this building by Lieut. Stern and other members the 'tank' committee in accordance with an intimation from the minister's private secretary.
>
> I am now asked whether this staff is working as a committee merely housed in this building or whether they are considered to be part of our organisation and their correspondence handled accordingly.
>
> I do not know whether any recent change has taken place in

the duties or composition of this committee which has been sitting at the Admiralty and I believe has, or had War Office representatives upon it, but so far as I am aware no representative of the ministry.

Colonel Lee will remember consulting with director-general of munitions supply some short time ago when an inquiry was made as to whether the supply of the mechanisms which the committee was experimenting with could be taken over with advantage by this department of the ministry.

After a joint consultation and some talk with the secretary of the committee, the opinion was formed that the mechanisms in question had practically little or no affinity with our work.

I understand that the committee has expended large sums of money and we have not hitherto been associated with their work.

Is it now intended that this department shall assume definite responsibility for supply? If so arrangements will be proposed and the question of technical responsibility—*i.e.* whether the director-general of munitions design is associated with that side of the work—will need to be settled.

Sir Arthur Lee replied:—

I have discussed this curious situation with Lieut. Stern, and he has sent me the attached copy of the 'charter' which the minister handed to him. It apparently relieves you of any responsibility with regard to the business of the 'tank committee.'

To this Sir Frederick Black wrote:—

In order that effect may be given to the minister's instructions in regard to facilities for supply of materials, I propose to issue a confidential office memorandum to deputy directors-general and directors.

On Saturday, February 12th (the day on which the charter was signed), all preparations having already been made, orders were telephoned and telegraphed to Messrs. W. Foster & Company, of Lincoln, and the Metropolitan Carriage Wagon and Finance Company to start the production of 100 machines. Orders were also given to Mr. P. Martin of the Daimler Company to supply 120 engines by the end of June. Complete drawings and other details were available, and a contract was arranged on Tuesday, February 15th, at Birmingham.

Messrs. Beardmore, Messrs. Vickers, Ltd., and Messrs. Cammell Laird

& Company, agreed to produce the necessary armour-plate. Owing to transport difficulties in Glasgow, Squadron 20 were instructed to send at once two light lorries with drivers to assist the armour plate firms.

Although the order was placed for tanks which were to be proof against both German "S" and armour-piercing bullets, the committee had already in view a tank which should also be proof against armour-piercing shell. On February 14th, the following letter was sent to the War Office:—

> I write to say that the tank supply committee, as recommended by the conference, with some slight modification, has been constituted, the minister of munitions having signed the charter for its constitution under him on Saturday, February 12th, when informed that the War Office were writing to demand the supply of 100 machines. Orders for the engines and some other parts of the machines were sent out by telegraph and telephone the same day.
>
> 2. The tank supply committee, in addition to proceeding at once with the construction of the 100 machines ordered according to the sample inspected, has at its disposal a considerable sum of money for experimental work which will be carried on separate from the construction of the 100 machines, and without in any way delaying their production.
>
> 3. As you know, the machine approved and now being produced is furnished with bullet-proof protection alone. The tank supply committee, however, propose to try and evolve another and superior type of machine, and the lines upon which their experiments are going are the following:—
>
> To produce a tank which will not only be bullet-proof, but will be armoured so as to be proof against the high explosive shell from German field-guns, and also the projectiles fired by the small calibre quick-firing artillery which it is believed the enemy may bring against them.
>
> 4. So far as can be seen at present, to fulfil this requirement will necessitate thicker armour or a double skin of armour, which will largely increase the weight to be carried. This may or may not mean a larger machine, but it will entail engines of greater power than those used at present. It is doubtful, however, if the climbing capabilities of the machine, or its speed, will be much increased. It is not known whether the improved machine should carry 6-pounder guns as the present type, or whether

an attempt should be made to carry, say, field-guns, or even something larger.

5. Since this heavier machine is in its embryonic state and the ideas are at present entirely fluid, it will be of great assistance to the tank supply committee to know upon what lines the general staff consider the development of a superior machine should proceed. The following heads for this information are suggested:—

(a) The nature of the attack against which the armour is to be constructed.
(b) The armament which the machine should carry.
(c) The speed to be attained.
(d) The climbing powers.

All the above subject to the conditions that they are mechanical possibilities.

The War Office replied:—

I am directed to send you replies to the questions contained in the memorandum as follows:—

Q. I. The nature of the attack which the armour should be capable of resisting?
A. 1. Field-gun fire.
Q. 2. The armament which the machine should carry?
A. 2. No increase on present pattern.
Q. 3. The speed to be attained?
A. 3. Top speed six miles per hour.
Q. 4. The climbing power?
A. 4. To be capable of crossing a ditch 10 to 12 feet wide with a parapet 6 feet high and trench 4 feet 6 inches wide on the far side.

I am to add that it is important that the machine should not be increased in size to any great extent and the height should be kept down. If the reply to Q. 3 and 4 involves any large increase in size the additional speed, etc., should be dropped.'

The first report of the tank supply department, dated Monday, February 28th, sixteen days after it was created, gives some idea of how the business was tackled.

February 15th.—Offices in Hotel Metropole occupied.
February 16th.—First committee meeting for construction of

BULLOCK TRACK MACHINE

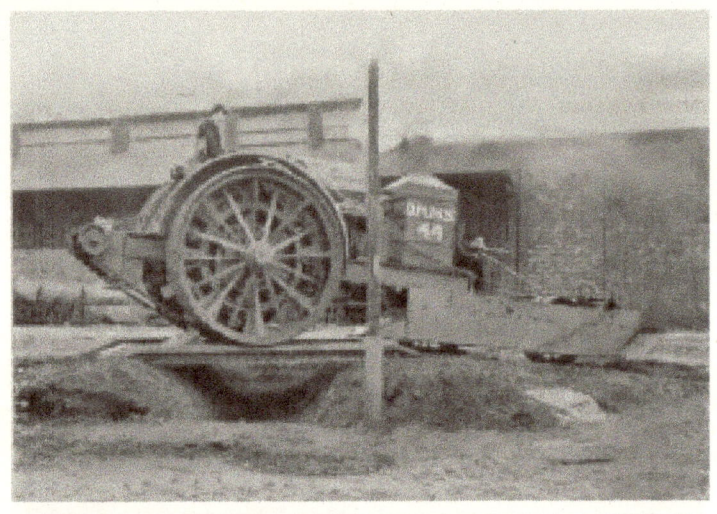

TRITTON'S TRENCH CROSSING MACHINE

100 tanks and experimental work for type of machine embodying improvements laid down by the War Office.

Main Contracts

Tanks.—The construction of 100 tanks has begun under arrangements with Messrs. Foster & Company, Ltd., of Lincoln (25), and the Metropolitan Carriage Wagon and Finance Company, Ltd. (75).

Armour Plates.—Orders have been given to Messrs. Cammell Laird & Company, Ltd., Sheffield, Messrs. Vickers, Ltd., Sheffield, and Messrs. Beardmore & Company, Ltd., Glasgow. Plates have already been rolled and are in course of transit for machining.

Guns.—Negotiations are in progress for supply of 200 6-pounder Q.F. guns by Admiralty working in conjunction with the deputy director general (D).

Machine-guns.—400 Hotchkiss machine-guns have been asked for. Decision as to stocks and clips still to be decided. Anticipate no difficulties.

Recoil Mountings for Gun Shields.—Negotiations for the manufacture of these are in progress.

Ammunition.—100,000 rounds 6-pounder ammunition required (85,000 rounds high explosive reduced charge and 15,000 case shot). Drawings have been handed to deputy director-general (A) Department for high explosive shells: drawings of case shot not yet to hand. Anticipate great difficulties.

Periscopes.—Two suitable periscopes and telescopes for guns are under construction for early delivery.

Lieut. Symes had been made responsible for the supply and inspection of all armour-plate; the Admiralty agreed to lend their inspector to overlook the construction of the Daimler engines; Lieut. Wilson agreed that he would be able to supervise the work of the Metropolitan Company at Wednesbury and of Messrs. Foster & Company of Lincoln, and Mr. F. Skeens was lent to the committee from the Admiralty by Mr. d'Eyncourt for armament work, and later elected a member of the committee.

The Admiralty now found that they were unable to provide more than 100 guns of the 200 promised, and an order for 100 guns was given to Messrs. Armstrong, Whitworth & Company; 100,000 shells were also ordered, and after a discussion whether black powder or high explosive should be used, it was decided to use black powder.

The use of case shot which we had suggested was turned down by the War Office, and this is interesting, since it was again proposed some twelve months later, was adopted and found most satisfactory in the battles of 1918.

Colonel Wheeler reported that Madsen guns would not be available, but it would be possible to supply Hotchkiss and Lewis guns. With these changes in our first suggestions we went on with the work.

Early in March, Colonel Swinton was appointed to the command of the corps which was to man the tanks. It was to be part of the Motor Machine-Gun Corps. Later on, it changed its name to Heavy Branch, Machine-Gun Corps, and still later to Tank Corps. I had held a commission as Lieutenant, R.N.V.R., but was transferred to the Motor Machine-gun Corps as a major on March 6th, 1916. In fact, I was very kindly given the first commission by Colonel Swinton in his new corps. Shortly afterwards. Major Wilson, Captain Symes and Lieut. Rendle were appointed and placed under my orders. The reason for calling the corps Heavy Branch, M.G.C., was to deceive the enemy and the inquisitive.

A camp was taken near Bisley, and Lieut.-Colonel Bradley, D.S.O., put in charge. Most of the officers and men who were drafted had but a very vague idea of what they had to do, and the whole of the Heavy Branch was generally known as the "hush-hush" crowd.

Colonel Swinton and Colonel Bradley spent days going round to the various O.T.C.s all over the country picking out young officers with the necessary qualifications to act as tank commanders, and no men were taken who had not a good experience of motors.

The original organisation was authorised to consist of fifteen companies. Each company was to have two sections of six tanks each, and the strength of a company to be fifteen officers and 106 men.

Colonel Swinton's intention was that as soon as the men had done a certain amount of elementary training, he would form three battalions, each consisting of five companies, and it was with this idea that all the preliminary training was carried out. In this, as in many other matters in the early days of tanks, Colonel Swinton's first views have been proved right by subsequent events. The organisation which was in use in the Tank Corps from January 1917, was battalion organisation, each battalion consisting of three companies of twenty-five tanks, but Colonel Swinton's proposed arrangement was rejected by G.H.Q., France, as soon as the establishment was sent out by the War

Office for their approval. They stated that they did not want battalions, but that the company was to be the tactical unit, and that companies must consist of twenty-five tanks each.

This meant altering the whole of the organisation on which the corps had worked at Bisley. Each company consisted of four sections, each under the command of a captain, with six subalterns, commanding the six tanks in the section. The total strength of a company was 28 officers and 255 other ranks. The companies were commanded by Majors Tibbetts, McLlennan, Holford Walker, Summers, Nutt and Kyngdon.

From March until about the middle of June all training was done at Bisley. It consisted, in addition to the ordinary recruits' training, of machine-gun and 6-pounder work, and the navy was very helpful in allowing a considerable number of both officers and men to be put through special courses on the 6-pounders at the Naval Gunnery School at Whale Island.

The question of protecting the tank against shell fire was now taken up, and a double skin was tried. It was discovered that a 3/8-inch plate, with half the metal stamped out, giving it the weight of 3/16-inch plate, and placed one foot in front of the ordinary armour plate of the tank, would detonate a German high explosive shell and prevent any damage. These experiments, however, proved of no practical value for the tanks, owing to the difficulties of construction.

In the middle of March Mr. Glynn West, controller of the shell department, was unable to obtain the regulation steel for the common pointed shell. He proposed using carbon steel, which is of slightly less penetrating power, and this was approved.

Colonel Swinton at this time ordered a full-sized model of a tank mounted on a rocking platform, in order to train his men to the peculiar motion of the tank. This was built by the department but never used.

For the object of guiding the tank into action small signalling balloons, the shape of observation balloons, were ordered.

Magnetos for the engines were ordered from America. This was the only part of the tank which was not English, and later on, when the home industry was developed, the magnetos also were made in England.

Prismatic peep-holes on the tanks were fitted, but these were later discarded owing to the danger from breaking glass.

On the 3rd of April the order for 100 tanks was increased to 150,

50 to carry 6-pounder guns and 100 to carry machine-guns, and a week later the order was changed to 75 of each armament. Tests were now started by Lieut. Symes with a German field-gun on a 2-inch high tensile plate.

The committee took up other ideas besides tanks. It designed a mechanical carriage for 5-inch howitzers and 60-pounders, a carriage to take the gun, its ammunition and its crew, and its design was approved in April. It also built an experimental plough for laying telephone wires. One of our reasons for working on other things besides tanks will be seen later.

In France they were most anxious for the coming of the tanks, and a letter, dated April 26th, from Colonel Swinton to General Butler, informs Sir Douglas Haig of the dates for delivery.

My Dear General,

Your letter of the 24th instant reached me today. Thanks for the promise of the R.E. officer. I am looking forward to his arrival.

About deliver, I know how anxious the commander-in-chief and you are to get some machines at an early date, and all of us here are equally anxious to expedite things in every possible way.

When I saw Sir Douglas Haig on Friday, the 14th, the idea was if possible to have some machines over in advance by the middle of June (not the 1st). I said that I feared it was not possible, but I deferred giving a final answer until I saw Stern, who had the manufacturers' progress charts.

On Monday, the 17th, when you saw Stern and me together we said that to get any machines over by the middle of June would not be possible, but that we were doing all we could to shove on with the production.

In regard to this matter, upon which so much depends, it is best to be categorical as to what we expect can or cannot be done, and so to avoid disappointment and the reversal of plans.

By 1st June.

No machines will be ready and no crews.

By 1st July.

Some practically finished machines will have been delivered at home which will be in a fit state to move and so to instruct men to drive, but owing to their design they will not be fit to take the field, even if they are manned by machine-guns and armed with M.G. supplied in France.

In regard to this I shall be able to give more definite information in four weeks' time.

By 1st August.

The supply committee informs me that all the machines will be ready and some will already have been shipped to France—strikes and acts of God excepted.

The number of crews that will be trained will depend on the rate at which the machines are received during July, but I anticipate that crews for seventy-five 'tanks' will be fully trained in any case.

Stern leaves tonight for a tour of the works, after which we shall be in possession of more positive information, but it will certainly not alter the position on June 1st or July 1st.

As regards the trial, the War Office has just arranged to take up our ground at Thetford, and is arranging for troops to be sent there to dig our manoeuvre ground. I anticipate that by the middle of June, or the latter half, we shall have some partially finished tanks down there which have been rushed there as soon as they are capable of movement, to train men to drive. As soon as there is anything to see there I will give ample warning, so that you can come over and frame your own ideas on the things.

I am afraid this letter does not contain what you would have liked to hear, but it is the cold truth and shows the real situation.

Other people were also very anxious to obtain tanks—but not the kind we were building. The secret of our work was very well kept in the Ministry of Munitions, not even the inquiry office being in possession of the true facts. This had its disadvantages, however, and caused us unnecessary work, for very frequently we had inquiries from enthusiastic manufacturers of gas, oil and water tanks, who were anxious to secure orders in their own particular lines.

On one occasion a staff officer at the War Office rang us up and asked if we were the "tank" department. On being told that we were, he asked when delivery of his oil tanks might be expected. He was politely informed that we could not tell him, as we were not building oil tanks. He then asked what sort of tanks we were interested in—gas or water—and on receiving the reply that we were interested in neither, he got very much annoyed and banged his telephone-receiver down.

105 H.P. Tractor

Major-General E. D. Swinton D.S.O.

On May 15th it was decided that the tanks should be numbered, the 6-pounder tanks from 500 to 574, the machine-gun tanks from 500 to 574, and that, as a disguise, all should have painted on them in Russian characters, "With care to Petrograd."

Two experiments were now made to protect the roof of the tanks against bombs, one with splinter proof mattresses such as are used on battleships, the other with expanded metal. As a matter of fact, in the first battles a wood and wire penthouse roof was used, and thought to be unnecessary. After that no special device was tried.

Luminous tape was prepared which, laid on the ground, was to guide the tanks to their positions at night. The men who acted as tank guides were also provided with electric lights on their backs, red and green, by means of which they signalled at night to the driver which way to turn.

The tank supply department also supplied the Tank Corps with its tractors and workshop wagons, which were specially designed for them and built by Messrs. Foster of Lincoln.

On June 5th it was decided to paint the tanks light grey.

We now had tanks available for training the men of the corps, and Colonel Swinton had succeeded in getting lent to him a part of Lord Iveagh's estate at Thetford, in Norfolk. A large number of men of the Royal Engineers were sent there and dug an exact copy of a part of the line in France, and the Royal Defence Corps sent up two battalions who guarded every entrance to the ground, which was about five miles square. No one was allowed to go in without a pass, and prominent notices explained to the public that it was a dangerous explosive area.

A story was current at the time among the local population that an enormous shaft was being dug from which a tunnel was to be made to Germany. Here in the middle of June two of the six companies arrived and began to train with six tanks each. This training went on until the middle of August. Several displays were given there during the summer, and live 6-pounder shells were used. The King, Mr. Lloyd George and Sir William Robertson were among those who saw our displays, and in June Colonel Estienne, who later on was to command the French tanks, visited the camp.

There was great difficulty and delay in making the 6-pounder armour-piercing shell, and we finally discovered that the Japanese had some 25,000 shells which were originally made by Armstrongs. These were shipped back to England for our use.

On June 5th I saw General du Cane, Director-General of Munitions Design, and General Headlam with reference to the new 60-pounder gun-carrying tank, and it was decided that modified drawings should be pressed forward with all haste. We were unable, at the time, to get any further orders for tanks, and we wanted to keep our works busy and avoid discontinuity of production.

Communication with a tank was one of the greatest troubles, and at this time experiments were made with a daylight signalling lamp, with wireless and with semaphores.

On June 10th it was decided to design and build a tank capable of resisting field-guns. Mr. Tritton had already got out certain designs, and experiments were carried out at Shoebury with Beardmore plates of 1 inch, 1½ inches, and 2 inches thickness. An order was given to the Daimler Company to construct at once a double 105 h.p. engine for this heavy tank. It was, however, never completed. Mobility was thought to be a surer defence than heavy armour.

On June 19th a model of the new gun-carrying machine was placed before the ordnance committee and its principle explained. . Major Wilson and Major Greg of the Metropolitan Company were also present. Instructions were from the minister, Mr. Lloyd George, on the 16th of June to proceed with the construction of fifty of this type of machine. A committee was formed to deal with this question. It consisted of Major Stern (Chairman), Mr. d'Eyncourt, Colonel Evans, Major Wilson, Captain Symes, Lieut. Holden and Major Dryer (representing the ordnance committee). Colonel Goold Adams, director-general of munitions inventions, agreed to help.

Mr. Norman Holden, who had been invalided out of the service after being severely wounded in the armoured car attack on the Peninsula of Gallipoli, joined my staff as my deputy at this time.

Lieut.-Colonel Solomon J. Solomon, R.E., now undertook to camouflage the tanks, and was supplied with several tons of paint.

About this time Sir Arthur Conan Doyle was writing to the press and pointing out that unnecessary casualties were caused by making frontal attacks on German machine-guns with unprotected infantry. He suggested that light armour should be worn, and that the authorities were wasting lives by not using it.

Mr. Montagu asked me to see him and to show him that we were doing something still better to protect the infantry by mechanical means from mechanical guns. He was very much interested in our developments.

From that time I kept in close touch with him, knowing his great knowledge of the history of war. I told him that our idea was that once we had tanks in large numbers we could bring back the element of surprise which was now entirely lacking in the attack. Although he believed in mechanical warfare, he doubted this. He doubted it until the battle of Cambrai in November 1917, when he wrote to me:—

> Windlesham,
> Crowborough,
> Sussex,
> November 22nd.
>
> My Dear Stern,
> I think your tactical ideas have been brilliantly vindicated by this battle, and that you should have warm congratulations from all who know the facts.
>
> Yours very truly,
> A. Conan Doyle.

It was rumoured at this time, also, that information was leaking out from Birmingham, and twelve men and one woman, who were working for a Swiss company at the Metropolitan Works, were closely watched. One of the men wished to return to Switzerland, but was interned.

I was pressing all the time for further orders, but on the 10th of July a letter was sent to the director of staff duties by Brigadier-General Burnett Stuart from G.H.Q., France, asking that further orders should be delayed. He wrote:—

> It is hardly possible to decide now, with the knowledge at our disposal, whether more tanks should be ordered or the type changed.
> Before any judgement can be formed it will be necessary to see at least twenty Tanks fully equipped and manned, functioning in accordance with some definite tactical scheme. It will also be necessary to view the French experiments, which they have informed us that they propose to hold shortly with their tanks.
> Can you say, please, for how long a decision may be deferred without endangering the continuity of manufacture?

This letter was sent on to Colonel Swinton, who wrote back on July 12th:—

> The reply to the question in the last paragraph of the letter from G.H.Q., France, which was enclosed, is as follows: the de-

cision as to a further supply of tanks, if it is affirmative, should be immediate.

As regards the manufacture it is a question of engines, guns, gun mountings, gun ammunition and various small parts. The absolute continuity of supply is, as a matter of fact, already broken, but so far the skilled men have not been dispersed.

There is one other point in regard to the matter to which I have not made any allusion, and that is the question of the provision of personnel. So far the heavy section has been able to obtain a very good class of man—one quite above the average—but I rather imagine that the source of supply is to some extent exhausted. In the event of the supply of more tanks, therefore, we cannot count on obtaining men of the stamp necessary from the open market, and to man a special corps, as the heavy section is, with the personnel of inferior quality would be fatal. The work needs men of some education, a mechanical bent, good physique and intelligence.

In the event of the development of this branch being required I would suggest that the most satisfactory way of obtaining the personnel would be to transfer the personnel of some existing unit which is trained in somewhat similar duties to the heavy section; for instance, men of the Royal Marine Artillery would, I think, be eminently suitable for duty with the heavy section, as they are trained in gunnery, machine-gunnery and in machinery. I refer to this subject because it is germane to that of time. If more tanks are to be constructed it will be essential to take the requisite number of men *en bloc* from some existing formation, and that at the same time as it is decided to increase the heavy section.

Further I would add that an early decision is necessary, because any increase in the number of tanks will necessitate a tremendous expansion of the ancillary services connected with the maintenance and repair of the unit, the importance of which we are only just beginning to be in a position to gauge.

About this time I found it very difficult, as a supply department, to work with a committee, some of whom wished every point referred to them. After reflection I decided that the only way to work successfully was to turn it into an advisory committee. At a meeting of the committee I proposed this. I explained that I and my department could alone be held responsible on questions of supply, that the different members of the committee had been appointed for their very special knowledge on particular subjects and had given invaluable advice individually, and that it was the name of "supply committee" alone that had caused any misunderstanding as to the duties of its members. The resolution was carried unanimously and was approved by the minister of munitions, Mr. Montagu. I was then appointed director of the tank supply department, with Lieut. Norman Holden as my assistant, and the powers and duties of the committee were transferred to me. Mr. d'Eyncourt agreed to act as chief adviser on all technical and experimental matters, and Mr. P. Dale Bussell to assist in matters of general organisation and procedure.

At the end of July we were told that the War Office proposed to send a few tanks out to France at once. They were, however, in such a state of repair that it would take two months at least to get them ready.

I immediately went down with Colonel Sykes to the repair shop unit of the heavy branch of the Machine-Gun Corps at Thetford and found that it was totally inadequate to cope with the work of tuning up the tanks in a hurry. In those days the machinery and skill needed for taking off the track were not developed in any way.

I returned to the War Office and told them that I would guarantee to have all the machines put in order within ten days. Then I went to Birmingham and asked for volunteers from the employees of the Metropolitan Carriage Wagon and Finance Company to get the tanks ready within a week to go to France.

I told them that the accommodation and food would be difficult to find, but without the slightest hesitation Mr. Wirrick of the Metropolitan Company and forty men started for Thetford. They were billeted by the chief constable. The difficulty was the food. The army could not supply it. I therefore went to Colonel Thornton, general manager of the Great Eastern Railway, and he immediately put a restaurant car on a siding at the camp, and fed the men until the work was done. It took them less than ten days.

This is only one of the instances of the magnificent patriotism and unselfishness of the industrial workers, who were ready to labour night and day for the tanks, from the making of the first experimental machine until the Armistice was signed in November 1918.

Meanwhile, Mr. d'Eyncourt and I saw Sir William Robertson. We were most anxious that the tanks should not be used until they had been produced in large numbers. We urged him to wait until the spring of 1917, when large numbers would be ready. I also wrote to the Minister of Munitions.

August 3rd, 1918.

I beg to refer to our conversation regarding the order for 150 tanks. My department was originally given an order to produce 150 tanks with necessary spares, and I was under the impression that these would not be used until the order had been completed, therefore the spares would not, in the ordinary way, be available until the 150 machines were completed.

From the conversations I have had with Mr. Lloyd George and General Sir William Robertson, and information received from Colonel Swinton, I believe it is intended to send small numbers of these machines out at the earliest possible date, and I beg to inform you that the machines cannot be equipped to my satisfaction before the 1st of September. I have therefore made arrangements that 100 machines shall be completed in every detail, together with the necessary spares, by the 1st of September. This is from the designer's and manufacturer's point of view, which I represent.

I may add that in my opinion the sending out of partially equipped machines, as now suggested, is courting disaster.

I had seen Mr. Lloyd George, the Secretary of State for War, and he heartily agreed with me, but on the other side it was urged that the heavy casualties in the Somme offensive of July 1st, the want of success against the German lines since then, and the approach of winter without any appreciable advance having been made, all tended to lower the *morale* of the troops, and it might therefore be necessary to use these new weapons in order to raise it again. Our reasons for desiring to wait until the spring were understood, but we must be prepared to throw everything we had into the scale. The slightest holding back of any of our resources might, at the critical moment, make the difference between defeat and victory. Such were the arguments used.

The French also had begged us to delay until their own tanks were ready for action. M. J. L. Breton, the under-secretary of state for inventions, had pressed M. Thomas, their minister of munitions, to build tanks in large numbers; he was very anxious that the French should share in the first surprise, and, when this was impossible, urged that they should continue to build as rapidly as possible. It was on September 28th, just after the first tank battle, that he wrote:—

I think it is now unnecessary to labour the imperative need for pressing forward with the construction of our offensive caterpillar machines as quickly as ever it is possible to do so.

The English, by using prematurely the engines which, to their credit, they constructed much more rapidly than ourselves, have debarred us of the use of the element of surprise, which should have enabled us easily to pierce the enemy's lines, though they have more or less rendered us the service of convincing even the most sceptical and most red-tape bound.

He urged his views again in a letter of October 20th:—

It seems to me more than ever indispensable to take steps towards pushing on with the construction of more powerful machine, better armed, and, above all, more heavily armoured.

The heart-breaking precipitation of the English in prematurely using their machines, before we were in a condition to deliver to the enemy the decisive blow—which putting into the line several hundred of our machines would have enabled us to do—unfortunately no longer allowed us to anticipate the effect of the element of surprise, which would have been irresistible.

At the beginning of September, Mr. Montagu saw Sir Douglas Haig himself, and found him most sympathetic in hearing the views of those who were working and thinking and inventing at home, but he held out little hope of keeping the tanks until the spring. They would have to be used that autumn and used soon.

So it was arranged. The tanks at Thetford were entrained at night and taken by rail to Avonmouth. There they were shipped to Havre, taken to a village near Abbeville and from there sent up to a point fifteen miles behind the line. Moving tanks was in those days a very difficult business. The sponsors, each weighing 35 cwt. (gun included), had to be unbolted and put on separate trucks, and in that journey from Thetford to the front tanks and sponsors were loaded and unloaded five times. The first party of the men of the Heavy Machine-Gun Corps crossed to France on August 18th. Other parties followed, and on September 15th, seven months after the first order was given by Mr. Lloyd George, the tanks went into action.

CHAPTER 5

Tanks in Action
SEPTEMBER 1916 TO OCTOBER 1916

The tanks were already in France and waiting to go into battle, but the secret had been well kept, how well was shown by a thing that happened on the very morning in September when I was leaving for the Somme for the first tank action.

A civil servant, an assistant secretary, came to see me on this eventful morning just as I was starting.

He told me that as my department was of no real importance, since he had no knowledge what it was, he had arranged that during the next Sunday all my papers and drawings were to be moved out into a small flat in a back street opposite the Hotel Metropole.

This was no time to argue; my train left in a few minutes; once more the famous Squadron 20 to the rescue. I told him that the department could not move, as it was concerned in matters of the greatest national importance, and would require before long a very large building of its own. This had no effect on him, so I gave instructions to one of my officers in his presence to put an armed guard on my office while I was away, and to resist any attack. Should he make an attempt he was to be arrested, taken to Squadron 20's headquarters at Wembley, tied to a stake for twenty-four hours and the reason carefully explained to all and sundry, especially newspaper reporters.

Fortunately for him no attempt was made, but on my return we were offered, amongst other buildings, the Colonial Institute and the Union Club. Finally, we took Nos. 14, 17 and 19, Cockspur Street, and even these blocks of buildings proved too small.

I arrived late at night on September 16th at Beauquesne, advanced headquarters, and found that an old friend of mine, Major A. H. Wood, was town major. Here I met Colonel Elles, who originally came to Hatfield for the B.E.F., and from him I learnt of the great victory of the tanks the day before.

In this, their first battle, forty-nine tanks were used, but of these seventeen did not reach their starting-point. They were either ditched on the way or broke down through mechanical trouble. The ground over which the remaining thirty-two attacked had been heavily "crumped" in places, but the weather had been fine and dry, so that the ground was not unfavourable. The tactical idea for their use was that they should work in sub-sections of two or three machines against strong points.

Every tank was given the route that it was to follow, and the time that it was to leave the starting-point. In most cases this was half an hour before zero, which was fixed for dawn, so that the tanks should reach the German trenches five minutes ahead of the infantry.

The risk of these tactics was that the tanks, by starting before the infantry, might prematurely draw the enemy's fire, but this risk it was decided to take. When our own barrage came down on the enemy's front line it left lanes free from fire, and by these lanes the tanks were to advance.

The seventeen that worked with the 15th Corps were the most successful. Their starting-points were round Delville Wood, and eleven of them crossed the German trenches. One gave great help to the infantry when held up by wire and machine-gun fire before Flers itself. Its commander put it across the trench, which he raked with his fire, then, travelling along behind the trench, he captured 800 prisoners.

Another tank pushed into Gueudecourt and attacked with its 6-pounders a German field battery of the same calibre as our 18-pounders. It destroyed one gun, and was then hit and caught fire. Only two of its crew got back, but the total casualties of the tank crews were very small.

Altogether, of the thirty-two which reached their starting-points, nine went ahead of the infantry, causing great loss to the enemy; nine more, though the infantry got ahead of them, did good work in demolishing strong points where the enemy still held out. The remaining fourteen broke down or were ditched.[1]

On Sunday, the 17th, Sir Douglas Haig appeared in front of General Butler's offices and congratulated Colonel Swinton and me. He

---

1. I have an interesting relic of the battles of the tanks on the Somme. See Mr. Ashmead Bartlett's letter in Appendix 1.

"Little Willie" Machine

"Mother" the original Tank

said, "We have had the greatest victory since the battle of the Marne. We have taken more prisoners and more territory, with comparatively few casualties. This is due to the tanks. Wherever the tanks advanced we took our objectives, and where they did not advance we failed to take our objectives." He added: "Colonel Swinton, you shall be head of the Tank Corps; Major Stern, you shall be head of the construction of tanks. Go back and make as many more tanks as you can. We thank you."

A few days later Mr. d'Eyncourt and I were received by Sir Douglas Haig. He again said, "Go home and build as many tanks as you can, subject to not interfering with the output of aircraft and of railway trucks and locomotives, of which we are in great need."

At last our contention had been proved. We had always been convinced that mechanics applied to war would save life, just as mechanics applied to industry saved labour; that since there were limits to human endurance we must use steel instead of flesh and muscles, and that the only way to meet the machine-gun was with armour-plate.

That day I motored to Amiens for lunch with Major Wood and went to Bray-en-Somme, where the tanks were parked at a place called "The Loop"—new tanks and battered tanks together. I met my brother, Major Stern, and Colonel Thynne, in command of a composite regiment of South Irish Horse and Wiltshire Yeomanry, and took them over to see the tanks, which had created an immense sensation. We met the cavalry returning from the battle front. They had not been used. The lesson had still to be learnt that until the tanks could overwhelm the machine-gunners there could be no chance for the cavalry.

I dined with General Butler at his mess, and left by car for Paris on Monday morning, September 18th, at 9 a.m., with General Butler and Colonel Swinton, to see what the French were doing in the way of tanks.

We arrived in Paris in time for lunch, saw the first French tank at Marly and General Estienne, the first Commander of the French Tank Corps (*Artillerie chars d'assaut*).

After dinner we motored to Boulogne, arriving about 2 a.m. There a destroyer was awaiting us. It was not supposed to leave until daybreak, but Captain Evans, of South Pole fame, took us on board at once, and we reached Folkestone within an hour. London we reached by car before nine o'clock on Tuesday morning.

At 10.30 a.m. I had a meeting at the War Office with the Secretary of State for War, Mr. Lloyd George, General Butler, General Whigham, and Dr. Addison, representing the Minister of Munitions (Mr. Montagu, the Minister of Munitions, was abroad at the time).

At this meeting General Butler said that Sir Douglas Haig required 1000 tanks to be constructed at once. After discussion it was decided that this should be done, and orders were immediately placed with the manufacturers. This order meant the building of 30,000 tons of armoured vehicles, besides at least 1000 6-pounder guns and 6,000 machine-guns.

The mere tactical record of what the tanks did at Flers and Gueudecourt gives no idea of the morale effect of the first appearance of this new and strange weapon. It astonished and terrified the enemy. It astonished, delighted and amused its friends. War correspondents vied with each other to find the vivid, unexpected word that would do justice to its half-terrible, half-comic strangeness (and yet give away no secrets), and the humorists of the battalions sharpened their wits on it. They communicated their gaiety, through their letters, to the people at home. The jolliest, most fantastic of them all was a letter from a soldier to his sweetheart, which appeared in the newspapers at the time. It could not be left out of a book on tanks.

> They can do up prisoners in bundles like straw-binders, and, in addition, have an adaptation of a printing machine, which enables them to catch the Huns, fold, count, and deliver them in quires, every thirteenth man being thrown out a little further than the others. The tanks can truss refractory prisoners like fowls prepared for cooking, while their equipment renders it possible for them to charge into a crowd of Huns, and by shooting out spokes like porcupine quills, carry off an opponent on each. Though 'stuck-up,' the prisoners are, needless to say, by no means proud of their position.
> They can chew up barbed wire and turn it into munitions. As they run they slash their tails and clear away trees, houses, howitzers, and anything else in the vicinity. They turn over on their backs and catch live shells in their caterpillar feet, and they can easily be adapted as submarines; in fact, most of them crossed the Channel in this guise. They loop the loop, travel forwards, sideways and backwards, not only with equal speed, but at the same time. They spin 'round like a top, only far more quickly, dig themselves in, bury themselves, scoop out a tunnel, and come out again ten miles away in half an hour.

A little later on I took Mr. Wells to Birmingham to show him how his idea had at last been realised. He wrote an article on what he saw, prophesying, as only he could, what would come of these new weapons, and urging that the factories should not be robbed of the men who could build them. At the time the article was forbidden by the censor. I will quote from it his description of the tanks. It was one of the earliest authentic descriptions written at a time when so much was appearing in print that was entertaining but untrue.

October, 1916.

Tanks

The young of even the most humble beasts have something piquant and engaging about them, and so I suppose it is in the way of things that the land iron-clad, which opens a new and more dreadful and destructive phase in the human folly of warfare, should appear first as if it were a joke. Never has any such thing so completely masked its wickedness under an appearance of genial silliness. The tank is a creature to which one naturally flings a pet name; the five or six I was shown wandering, rooting, and climbing over obstacles, round a large field near X——, were as amusing and disarming as a litter of lively young pigs.

In a little while there will probably be pictures of these things available for the public; in the meanwhile, I may perhaps give them a word of description. They are like large slugs; with an underside a little like the flattened rockers of a rocking horse; slugs between 20 and 40 feet long. They are like flat-sided slugs, slugs with spirit, who raise an inquiring snout, like the snout of a dogfish, into the air. They crawl upon their bellies in a way that would be tedious to describe to the inquiring specialist. They go over the ground with the sliding speed of active snails.

Behind them trail two wheels supporting a flimsy tail, wheels that strike one as incongruous as if a monster began kangaroo and ended doll's perambulator. These wheels annoy me. They are not steely monsters; they are painted the drab and unassuming colours that are fashionable in modem warfare, so that the armour seems rather like the integument of a rhinoceros. At the sides of the head project armoured cheeks, and from above these stick out guns that look very like stalked eyes. That is the general appearance of the contemporary tank.

It slides on the ground; the silly little wheels that so detract from the genial bestiality of its appearance dandle and bump behind it. It swings round about its axis. It comes to an obstacle—a low wall, let us say, or a heap of bricks—and sets to work to climb with its snout. It rears over the obstacle, it raises its straining belly, it overhangs more and more, and at last topples forward; it sways upon the heap, and then goes plunging downwards, sticking out the weak counter-poise of its wheeled tail. If it comes to a house or a tree or a wall, or such like obstruction, it rams against it so as to bring all its weight to bear upon it—it weighs *some* tons—and then climbs over the debris. I saw it, and incredulous soldiers of experience watched it at the same time, cross trenches and wallow amazingly through muddy exaggerations of shell holes. Then I repeated the tour inside.

Again the tank is like the slug. The slug, as every biological student knows, is unexpectedly complicated inside. The tank is as crowded with inward parts as a battleship. It is filled with engines, guns and ammunition, and in the interstices, men.

'You will smash your hat,' said Colonel Stern.

'No, keep it on, or else you will smash your head.'

Only Mr. C. R. W. Nevinson could do justice to the interior of a tank. You see a hand gripping something; you see the eyes and forehead of an engineer's face; you perceive that an overall blueishness beyond the engine is the back of another man. 'Don't hold that,' says someone. 'It is too hot. Hold on to that.' The engines roar, so loudly that I doubt whether one could hear guns without; the floor begins to slope and slopes until one seems to be at forty-five degrees or thereabouts; then the whole concern swings up, and sways and slants the other way. You have crossed a bank. You heel sideways, through the door, which has been left open, you see the little group of engineers, staff officers and naval men receding and falling away behind you.

You straighten up and go uphill. You halt and begin to rotate. Through the open door, the green field with its red walls, rows of worksheds and forests of chimneys in the background, begins a steady processional movement. The group of engineers and officers and naval men appears at the other side of the door and further off. Then comes a sprint downhill. You descend and stretch your legs.

About the field other tanks are doing their stunts. One is struggling in an apoplectic way in the mud pit with a cheek half buried. It noses its way out and on with an air of animal relief. They are like jokes by Heath Robinson. One forgets that these things have already saved the lives of many hundreds of our soldiers and smashed and defeated thousands of Germans.

Said one soldier to me: 'In the old attacks you used to see the British dead lying outside the machine-gun emplacements like birds outside a butt with a good shot inside. *Now*, these things walk through.'

That no time should be wasted and that both the French and we should get the best results, I decided to arrange for a conference in Paris at which all our engineers should meet all the French engineers engaged on mechanical warfare, and further that the whole party should go to the front to inspect our tanks and see the results of the battle. Mr. Lloyd George gave me a letter to his friend, M. Thomas, the French Minister of Munitions.

The development of mechanical warfare in France was due very largely to two men: on the military side, to General Estienne; on the civilian side, to M. J. L. Breton, Under-Secretary of State for Inventions.

General Estienne was a great believer in the small tank which could be quickly carried by lorry to the battle-field, for attack or counter-attack, and the first French tanks constructed by the two big French armament firms of Schneider and St. Chamond were, like our experimental machine at Lincoln, merely steel boxes placed on copies of the Holt track. When tried they did not prove a success. General Estienne then consulted M. Louis Renault, famous for the Renault motorcar, and he designed and produced in thousands a light tank which played a big part in Marshal Foch's great counter-attack in July, 1918. None of the French tanks were able to cross large trenches, such as our Tanks could cross (the Schneider and the St. Chamond, owing to their design, the Renault, owing to its small size); but in the battle of July, 1918, fought on ground where the Germans at the end of their advance had had little time to fortify themselves, they were not called upon to cross specially-prepared trenches such as the Hindenburg Line.

Besides General Estienne, with his great military experience and enthusiasm, mechanical warfare in France owed much to the officers

TANK MOUNTING RAILWAY TRUCK UNDER OWN POWER

TANKS ATTACK IN THIEPVAL, 1916

THE KIND OF GROUND OUR TANKS HAD TO WORK ON.
EVERYWHERE SHELL-POCKED, 1916

TANKS IN ACTION

under him, notably Commandant H. Michel, Capitaine de Poix, and, later, General Monhover and Capitaine Communeau.

Nor would the larger development of tanks in France ever have taken place had it not been for the broad-minded enthusiasm and intelligence of Colonel Aubertin, who was in charge of all tank matters at the French War Office, under M. Clemenceau, the Minister of War. He eliminated all "red tape" and worked most loyally and enthusiastically with ourselves and, later on, with the Americans also. His deputy, Capitaine Parville, reflected the qualities of his chief.

It was an undertaking to get everybody assembled and transported to the front, especially civilians, with all the restrictions imposed both in the French and English zones. However, before the end of the week (September 23rd) our party had started for Paris. It consisted of Mr. d'Eyncourt, Captain Holden, Commander McGrath, Captain Symes, Major Wilson, Mr. Skeens, and Major Buddicom; Mr. Tritton of Messrs. Foster; Mr. Brackenbury and Colonel Hadcock of Messrs. Armstrong, Whitworth; Mr. Searle, representing the Daimler Company; Mr. E. Squires, Major Greg and Mr. Stockton of the Metropolitan Carriage Wagon & Finance Company, and myself. The day before, the War Office agreed to my being made temporary lieut.-colonel, with authority to wear badges of rank on going to France.

We arrived in Paris and saw the French tanks. We met General Mouret and General Estienne. We met the representatives of the two French firms, M. Deloulle and M. Brillie of Schneider, and Dr. Laurens Dutilh and Colonel Remailho of St. Chamond. We visited the Schneider factory at St. Ouen near Paris.

It was very strange that at the conference between the English and French engineers at Marly, on the tank ground there, it was discovered that few of the Frenchmen could talk English and few of the English could talk French, but both could—up to a certain point—talk German, and it was by means of this language that they made each other understood.

On September 26th, very early in the morning, we set out by car, an Anglo-French party of thirty-five, for the Loop where the tanks were parked. At Amiens we lunched, and there I had enormous difficulties with the A.P.M. getting White Passes for the whole party. In the end we reached the Loop, inspected the tanks and saw one of them give an exhibition of crossing a deep road. I arrived back in London on Thursday, the 28th of September.

In order to get going on the big order, we had to increase the armour-plate capacity, and in addition to Beardmore, Cammell Laird and Vickers, we brought in Edgar Allen, Armstrong Whitworth, Firth's and Hadfields.

The chief difficulty in producing tanks in numbers was to get engines, and we at once took steps to find engines in America of the necessary horse-power. Our efforts were not successful, and an arrangement was made with the Daimler Company to supply us.

On October 16th a further 100 tanks were ordered to keep the factories going until the design for the 1000 had been settled.

The order for gun-carrying tanks was now reduced from fifty to forty-eight, the remaining two being made into labour-saving salvage machines for tanks, fitted with cranes on top.

In order to build the first tanks without any avoidable delay, it had been necessary to use only existing designs of engines and transmissions. Consequently, the transmission was not ideal, the steering being done by putting a brake on the differential of the old 105 h.p. tractor. When it was seen that the tanks were a success, I decided that every transmission that had any possible chance of success should be built experimentally. On the 3rd of October the orders were given. Mr. Tritton was to carry out an idea of his own, a double-engine tank (known as the "Whippet"), which was to be steered by accelerating one engine or the other; the Daimler Company were to construct a petrol electric transmission of their own; Mr. Merz was to design and build an ordinary electrical tramway transmission, with the British Westinghouse; Mr. Wilson, with Messrs. Vickers, was to build a Williams-Janney hydraulic transmission; the Hele-Shaw Company their hydraulic transmission; the Metropolitan Company Major Wilson's epicyclic and also Wilkin's system of multiple clutches; and the French St. Chamond Company arranged to have their petrol electric transmission fitted into a hull which I was to send them. This made eight in all.

The total order was now for 1250 tanks. It was an immense order to get placed. Large grants had to be made by the Treasury to extend the factories for the production of engines, guns, armour-plate, steel castings and other things. New factories had to be erected for the assembly of tanks.

It was about this time that I took Mr. Wells down to Birmingham, and he wrote, in the article from which I have already quoted,:

> I saw other things that day at X——. The tank is only a beginning in a new phase of warfare. Of these other things I may

only write on the most general terms. But though Tanks and their collaterals are being made upon a very considerable scale in X——, already I realised as I walked through gigantic forges as high and marvellous as cathedrals, and from workshed to workshed where gun-carriages, ammunition carts, and a hundred such things were flowing into existence with the swelling abundance of a river that flows out of a gorge, that as the demand for the new developments grew clear and strong, the resources of Britain are capable still of a tremendous response.

Then, on October 10th, I received an official instruction from the army council cancelling the order for 1000 Tanks.

All the manufacturers who had had any experience of the methods of the tank department up till then, had worked with the greatest enthusiasm. This sudden cancellation came as a thunderbolt. I immediately went to see Mr. Lloyd George, the Secretary of State for War. He said that he had heard nothing of the instruction. I told him that I had, with enormous difficulty, started swinging this huge weight, and that I could not possibly stop it now. I told him that he could cancel my appointment, but he could not possibly get me to cancel the orders I had placed. Sir William Robertson, the chief of the imperial general staff, then appeared, and Mr. Lloyd George said that he could not understand how this order could be cancelled without his knowledge, since he was President of the army council. He asked me to tell Sir William Robertson what I had told him. This I did. Excusing myself owing to pressure of work, I then left the room.

The order for the production of 1000 tanks was reinstated next day.

CHAPTER 6

Production on a Large Scale
October 1916 to April 1917

The business of the department now became more varied, and as the need for absolute secrecy no longer existed, our name of T,S. department (for up to this time, not even the word tank was used) disappeared. I was made director-general of the department of mechanical warfare supply, with Mr. P. Dale Bussell and Captain Holden as deputy directors-general. Sir E. H. Tennyson d'Eyncourt, K.C.B., and the Hon. Sir Charles Parsons, K.C.B., became chief technical advisers. Sir William Tritton was appointed director of construction, and Major W. G. Wilson, director of engineering.

Other changes had taken place. Colonel Swinton had returned to his duties at the committee of imperial defence. Colonel Elles now commanded the tanks at the front, and General Anley, who had commanded a brigade in the Mons retreat, now took command of them in England. He had a keen sense of humour. One day a bombastic lieut.-colonel of tanks came into my office when the general was there.

"When shall I get my tanks?" said the bombastic officer. "The Commander-in-Chief is awfully annoyed that I have not got any yet."

"This," said the general, turning to me, "reminds me of the fly on the elephant's trunk apologising for its weight."

It was General Anley who started the famous tank camp at Wool. It was splendidly organised, and the Tank Corps, under his command, became a fine, well-disciplined force. We who were building the tanks had every encouragement and help from him, and were all very sorry when he was appointed to a command in Egypt.

Complete secrecy had had grave disadvantages, but now that we

T. E. ESTIENNE

THE "SCHNEIDER" FRENCH TANK

THE "SAINT CHAMOND" FRENCH TANK

were beginning to be known, we found others that before we had escaped. One was a great increase in our correspondence. We received letters at this time from men of all nationalities all over the world, not only sending suggestions for the improvement of tanks but making claims that the writers were the inventors of them. This is one of the letters from a claimant. It was addressed to the King's Secretary, Christal Palace, London.

Thursday,
May 31st, 1917.

Sir,

I Wright Having discoved the first Tank Pattern now hused at the War and Dementions for huse I sent it to the Admarality wich brought it out i have not received enything for same Pattern and Dementions for huse I feel I should have reeived Something for same unles mistake as been made Patterns Advertsed for in Paper. God Save the King.

During these days, when we were just starting production on a large scale, we had rather a shock when we received, from the highest authority in the ministry, a minute which said:

> The opinion has been expressed that the time has been reached when new factories cannot generally be expected to begin production in time to be of service in this war, and that the building of further new factories should not therefore be sanctioned.

However, we did not allow this to interfere with our work, and Mr. Montagu, the minister of munitions, gave us every help. Like Mr. Lloyd George, he believed in mechanical warfare and was ready to fight for it.

I paid many visits to the front with General Butler to discuss the question of design, and in November a meeting was held with General Davidson of G.H.Q., General Anley and Colonel Elles, at which it was arranged that the first 150 machines which had been completed should be called Mark 1, and of the 100 which had been ordered to keep the factories going, fifty should be Mark 2, and the second fifty Mark 3. Mark 2 was to have no tail, but spuds on the track plates, and new cast-iron rollers. Mark 3 was to have thicker armour, but otherwise be the same as Mark 2. Mark 4, of which 1000 were to be built, was to have the Lewis gun instead of the Hotchkiss. Its petrol tank was to be at the back instead of inside. It was to have wider shoes, thicker armour all over, and the sponsons, which had hitherto been carried to the battlefield separately on trucks, were to be carried on the machine and made to swing in when travelling by railroad.

An officer of the Tank Corps, who had once been in charge of the Lewis Gun School at St. Omer, was responsible for the decision to use the Lewis gun instead of the Hotchkiss. He insisted on it against the advice of the experts in tanks, who knew that the vulnerability of the outer cover of the Lewis gun and the size of its barrel made it very unsuitable for using in a loophole. In the following year experience in the field proved us to have been right. The Tank Corps told us that they could not go into battle with the Lewis gun in the front loophole, and that until we could make the necessary alterations to put back the Hotchkiss gun, no tank actions could be fought.

I said at this meeting that the first heavy machine with 2-inch armour would be ready for trial towards the end of January, and if it were decided to proceed with the manufacture, the rest would be available about September, 1917.

I then brought forward a suggested type of light machine with two engines designed by Sir William Tritton. This was the Whippet. General Davidson and Colonel Elles agreed that it would probably be very useful. If any large quantity of these machines was required for next year it was necessary to order the engines at once.

I asked that an early decision should be given in order that I could obtain 5000 engines.

At this meeting, I also pointed out to General Davidson that we ought to have a member of the army council or some high official appointed to the committee to give decisions for the War Office, and it was agreed that when any important questions in connection with design, output or dates of delivery should arise, Lieut.-Colonel Stern, General Anley, Colonel Elles, a representative of the general staff, France, and a representative of the War Office should attend. A meeting could be convened by any one of the above representatives under the chairmanship of Colonel Stern. It was agreed also that Major Knothe should act as technical liaison officer between Colonel Elles, General Anley and Colonel Stern.

At a meeting on Saturday, November 25th, Mr. Martin of the Daimler Company undertook to deliver engines as follows:—

January	20 a week.
February	28 ,,
March	35 ,,
April	40 ,,

rising to sixty engines a week in May.

The question of responsibility for inspection and design was settled at a meeting with Major General Bingham, director-general of munitions design, Sir Sothern Holland, director-general of inspection of munitions, and Mr. E. Phipps, secretary of the ministry of munitions. In the usual way the inspection would have come under Sir Sothern Holland, and the design under Major General Bingham, but as the tanks were still in such an experimental stage, it was agreed, at the request of the War Office, that, for the present, all tanks, accessories and spares should be manufactured and tested by the mechanical warfare department of the Ministry of Munitions.

A suggestion that the army council should be consulted on specifications before they were approved, and that no change should be made without reference to them, was not adopted, as it was contrary to the principle of the division of functions between the Ministry of Munitions and the War Office, but it was agreed that all requirements and suggestions made by the army council should be incorporated in the designs at the earliest opportunity.

I was continually pressing for still larger orders for tanks, and in October I wrote to Mr. Montagu to say that we must now decide whether the order for 1000 tanks for the following year was sufficient.

I told him that the Russian, French and Belgian Governments had asked if the British Government could supply them with tanks, and that I thought that our capacity for production should be immediately increased to 150 tanks a week. I also said that many defences against tanks had been suggested, but that the real defence was another tank. we had to decide now whether every tank should not have a gun or sufficient power to smash tanks.

The Russian Government was not asking only if we could build tanks for it. I was pressed, at the same time, by the War Office to give plans of the tanks to Russia. I opposed this most strongly, and put it off from time to time by saying that I was too busy to get these drawings ready. I was convinced, by the nature of the questions asked, that the request really came from Germany. Russia could have no need for detailed drawings, either for offence or defence. She could not need them for offence, because even with the plans, she had not the means to build tanks. She could not need them for defence, firstly because the Germans had no tanks to use on the Eastern Front, and secondly, because if they had, we could give Russia sufficient information about defence against tanks without sending drawings of them. However, the intelligence department of the War Office was very insistent, so, in consultation with the minister of munitions and Sir Eustace d'Eyncourt, it was decided to give the War Office a child's drawing and incorrect details. I am convinced they found their way into the hands of the German general staff.

Now, and always, we had difficulties with the War Office about men. Although the officers in my department had volunteered for active service and some of them had been in action, although they were all doing work of the greatest military importance and paying continual visits to the front, yet it was very nearly impossible to get any military acknowledgment. Although Lieut. Holden, Lieut. Rendle and Lieut. Anderson were taking a leading part in the development of mechanical warfare, and were recommended by the minister for promotion on the 20th of September, 1916, Lieut. Holden only received promotion in October 1918, while Lieut. Rendle and Lieut. Anderson remained unpromoted to the end of the war.

This was not the only difficulty. In the early days we found it very hard to get any staff at all, for the army refused to allow us men of military age. It was very necessary, however, that we should secure the services of a good transport officer to superintend the transport of tanks from the manufacturers to tank headquarters in France, a man with business experience and a man of the world. I asked Mr. George Grossmith if he would undertake this work. He was over military age, but jumped at the idea of being able to help in any way and accepted at once. He was given a commission in the R.N.V.R. under the Admiralty, and did valuable work from the time of his joining up in November, 1916, till the date of the Armistice. Since he was an actor, many attacks were made on him by jealous people. It was on the occasion of one of these attacks that I was called to the Admiralty to explain what he was doing for my department. I told them, and his work was heartily approved. The official whom I saw sent for the file of papers relating to his commission. He told a clerk, who had been at the Admiralty some forty years, to look it up.

"Under what heading?" said the clerk.

"Ministry of Munitions," was the reply.

"Did you say Ministry of Musicians?" said this clerk of forty years' experience, looking very puzzled.

At a conference held at the War Office on November 23rd, 1916, at which Sir Douglas Haig was present, I raised the question of the immediate release from the colours of men required for the building of tanks. I said that I had handed in a list of the names of these men to the labour supply department, and that the matter was urgent. Mr. Montagu told me that the labour supply department of the Ministry of Munitions was responsible, and that he would ask Mr. Stephenson Kent to see to the speedy release of these men, numbering something like 300 in all. General Whigham, on behalf of the War Office, promised to expedite the matter. Sir Douglas Haig agreed that it was urgent.

The general conclusion reached at this conference was:—

1. Tanks are required in as large numbers as possible.
2. It is important to get as many as possible before May.
3. It is very important to consider and adopt improvements in design from time to time, but almost any design now is likely to be better than no tank.
4. It is highly desirable that no other supply should be interfered with. If it is necessary to do so in order to fulfil the requirements, G.H.Q. should be informed through the proper channels before any action is taken.

In September there had been a great technical controversy at the meeting with the French experts in Paris. The petrol electric transmission which was in operation in the St. Diamond machine attracted me very much, for it gave greater ease in changing speed, though at the price of greater weight. At this time all the experts were against me, but later in the year G.H.Q. made such urgent demands for tanks that, in order not to lose time, I gave orders to the Daimler Company on January 5th, 1917, for 600 sets of petrol electric gear with a low gear of three and a half miles an hour, and a changing speed gear of five miles an hour. The machine had not yet been tested, but this was to prevent any delay should the trial machine be a success.

On Friday, January 12th, the Daimler petrol electric machine was tested at climbing out of shell holes in competition with a Mark 1. 6-pounder machine. Both machines were loaded up to the full with ammunition and so forth, and the results clearly showed that the Daimler petrol electric was unable to pull out of the shell hole except by a succession of jolts, produced by bringing the brushes back to a neutral position, raising the engine up to 1800 revolutions and then suddenly shifting the brushes up to the most advantageous position. This resulted in a maximum current of from 900 to 1000 amperes, and all agreed that it was unsatisfactory. So after a great controversy and many tests the petrol electric machine was rejected as untrustworthy, and all orders were cancelled.

In January 1917 Major Uzielli was sent to me by General Elles to ask the exact dates for the delivery of the tanks. I said that the arrival of machines in France depended on so many things, that it was impossible to promise an exact date, but that from the first week in March we ought to have from twenty to thirty a week, which would take six days in passage. The inevitable delays were due to the difficulty in getting material, the possibility of breakdown in testing, and delay in transit.

I said that I was strongly of the opinion that it was exceedingly risky to make any arrangements for putting tanks into the line before we had a large supply at tank headquarters in France.

On January 29th, 1917, Squadron 20 undertook the testing of all machines before shipment to France.

Towards the end of 1916, Mr. d'Eyncourt and I were very much troubled about the future of mechanical warfare. G.H.Q., France, and the War Office had their hands full in these strenuous times, but we pioneers in mechanical warfare knew that, to develop it to its greatest extent, we must have the military views of the tactics of Tanks at least twelve months beforehand, in order to get the right design and production. We had been told, too, that there was a great probability that no more tanks would be required.

Having tried every other means, we at last went to the Prime Minister, Mr. Lloyd George, and handed him a memorandum explaining the development and present position of mechanical warfare.

We reminded him that on February 2nd, 1916, our first tank had passed all the tests laid down by the army council. The value of the tank had still remained in doubt until the battle of September 15th, but G.H.Q. had then unreservedly accepted this type of warfare. The main criticisms after that battle were that the tank got bellied, or stuck, and that it could not keep up with the infantry, so that either it had to start first, in which case the German barrage prevented the advance of the infantry, or if it started at the same time the infantry got ahead of it.

We explained the improvements which since then had been made. We pointed out that now the tanks were more easily transportable; that they were in a marked degree better at overcoming the most difficult wet and muddy ground; that experiments were being made and were now drawing to a conclusion for lighter loading, thicker armour-plate and ease of control, and that the machinery was much more trustworthy and durable.

Mr. Lloyd George then called a meeting of the War Cabinet, at which representatives of the imperial general staff and Sir Eustace d'Eyncourt and I were present. The Cabinet agreed to my suggestion that a meeting should be held at the War Office between representatives of the War Office of the English and French General Staffs, English and French commanders of tanks, and English and French designers of tanks, and that this meeting should take place after they had viewed the trials, with all the new types of tanks, which were to be carried out at our experimental ground at Oldbury early in March.

It was clear at this time that if mechanical warfare was to make any real progress, an engine bigger than the 105 h.p. would be necessary. Unfortunately, the aircraft production department were seizing every possible source of supply. They had the first claim. We were not allowed, for example, any aluminium for our engines. mechanical warfare, in fact, was not yet acknowledged as a necessity. I tried to persuade the Daimler Company to design a new engine, but they were already fully taken up with work for the air department. The problem before me was to have ready in advance a new engine of greater h.p., and to have it in sufficient quantities to meet all possible demands for an increase in the number of tanks.

I found a well-known designer in internal combustion engines, Mr. H. Ricardo, who under the directions of Mr. Bussell (my deputy) and with the assistance of my officers, found a certain number of gas-engine firms, which the air department had considered unsuitable and had rejected. These firms were got together under my presidency and agreed to work jointly and produce an engine to be specially designed by Mr. Ricardo, an engine of 150 h.p., using no aluminium or high tensile steel. As soon as Mr. Ricardo had got out the designs, I submitted them to Mr. Dugald Clark, one of the greatest authorities in the world on internal combustion engines, who considered, after examining the designs, that I was justified in ordering 700 of these engines before one had been constructed for test.

On February 5th, 1917, a conference was held in the room of the minister (Dr. Addison), at which were present:—

The Master General of the Ordnance;
General Sir David Henderson;
General Butler;

Mr. Percy Martin;
Sir W. Weir;
Mr. S. F. Edge;
Colonel Holden;
Colonel Foster;
Mr. Herbert and
Colonel Stern.

General Henderson asked that I should be prevented from employing five special firms in making 700 Ricardo engines, in anticipation of tanks which had not yet been ordered. I said that I had ordered these engines with foresight to prevent the shortage of engines for tanks such as they were now experiencing with aircraft. In spite of this, the committee approved resolution. However, I took no notice of it. We continued the building of the 700 engines, and in order not to stop the continuity of manufacture, I gave an order for another 700. The first of the initial 700 had not yet been tested, but we believed in them.

On March 3rd, the exhibition took place at Oldbury, Birmingham, on the ground belonging to Squadron 20, of all the experimental machines which I had ordered immediately after the Battle of the Somme, 1916. The full programme and the names of those present is given in Appendix 2.

The day after this display the meeting arranged by the War Cabinet was held at the War Office to discuss the tactics of the future design of tanks, and there were present representatives of the French general staff, War Office, French headquarters staff, the officer commanding English tanks, the officer commanding French tanks, M. Breton (representing the French Ministry of Inventions) Sir E. d'Eyncourt and myself.

At this conference the present value of mechanical warfare and the future value of improved machines were acknowledged by all the military authorities. The French feeling about it was expressed in a letter which M. Thomas, the French Minister of Munitions, wrote to Dr. Addison on March 10th:—

My Dear Minister and Colleague,
The Under-Secretary of State for Inventions whom you were good enough to invite to come to England in order to assist at the trials of the new type of tanks prepared by your technical department and at the conferences which were to follow the

trials, has given me an account both of the extreme interest of these trials, which you desired that representatives of France should witness, and of the very cordial reception which he had from you.

I have already had the opportunity of reading the report of the mission which I sent at your request to accompany the Under-Secretary of State for Inventions. I am extremely happy to note the brilliant success of the researches, as a result of which, very shortly, the British armies will be equipped with a weapon from which a great deal has already been expected, and I have the honour to thank you for the courtesy extended to the members of this mission by the British Government and by their British comrades and colleagues.

I appointed to take part in this mission, in addition to the technical officers, a certain number of the principal engineers of munition workshops. I note that they have greatly profited from the experiments of which they have been witnesses. I hope that the information which they may have been able to give to your technical officers may have proved equally valuable to them. I am delighted to see a continuance and an extension in the domain of experiment of that collaboration which, in one way or another, has existed from the very beginning.

I am particularly pleased that the Under-Secretary of State for Inventions should have had the opportunity of entering into personal relations with yourself and with the officers of your ministry who are specially concerned with research and invention.

You were good enough to point out to him certain directions in which researches in common might be made. From this co-operation I anticipate the happiest results.

Please accept the best assurance of my highest consideration.

Albert Thomas.

On the same day that the conference with the French was held, another conference at the War Office decided that there should be nine battalions of seventy-two tanks each, with another 352 as first reinforcements, making 1000 in all. The commander-in-chief also asked for 200 light tanks to be delivered by July 31st. I said that there was no possibility of delivering any tanks for this fighting season except the 1000 which had already been ordered, but that the question of the improvement in design and production must be considered at once if

RENAULT TANKS

Brigadier General H. J. Elles, C.B., D.S.O.

the large numbers which I saw would be required in the early part of 1918 were to be ready in time.

On March 12th, I wrote to Dr. Addison:—

As director-general of the department which has been responsible for the design and which has produced every tank, I have persistently opposed the premature employment of tanks this year. Also the employment of practice tanks, *i.e.*, Marks 1, 2 and 3 in action this year.

At the War Office meeting last Sunday, General Butler assured me that sixty machines of Mark 1, 2 and 3, which are being kept in France ready for action only as a temporary measure, and which are really practice machines, will be returned for training purposes as soon as they can be replaced by the delivery of Mark 4 machines. (Mark 1, 2 and 3 are practice machines, and Mark 4 is the new fighting machine; twenty-five of the above sixty machines are, I believe, being sent from Wool, and will be returned to Wool.)

I consider it more than unwise to use practice tanks in action under *any* circumstances. They have all the faults that necessitated the design of last year being altered to the present design of Mark 4. In addition, the training of the men is being delayed by this action. Their failure will undoubtedly ruin the confidence of the troops in the future of mechanical warfare.

For the sake of sixty machines, the whole future of thousands of tanks will be most unjustifiably prejudiced.

I wish also to point out that even when Mark 4 machines are delivered, taking mechanical warfare as an enterprise, it is a most uneconomical expenditure of our resources to use them before we have large numbers, and the necessary central workshops ready in operation.

As the war proceeds and our losses in men accumulate, necessity will force us more and more into labour-saving devices, which will take their place in warfare as they have in commerce.

All care should be taken to foster the development of this new weapon, with the greatest caution. However excellent the design of a ship, without a rudder she will be wrecked. I believe mechanical warfare, without prudence, will share a like fate.

At the beginning of March, Mr. Eu Tong Sen, a member of the Federal Council of the Malay States, offered £6000 for the purchase of a tank, which the army council gratefully accepted on behalf of His Majesty's Government. The tank selected was one built by Messrs. Foster & Company at Lincoln, and had a plate put on it with the following inscription:—

Presented to H. M. Government by Mr. Eu Tong Sen, member of the Federal Council of the Malay States, on March 10th, 1917.

All Chinese ships and boats, large or small, have a large "eye" painted at each side of the bow. The Chinese explanation of the custom is, "No have eyes, how can see?" It seemed only right that this "landship," also, should see, and accordingly an eye was painted on each side of its bow.

In due course it was sent overseas, and was first commanded by 2nd Lieut. J. M. Oke, and christened "Fly-Paper." Before the Cambrai battles in November 1917 it was re-christened "Fan-Tan," and in these battles was commanded by Lieut. H. A. Aldridge. It was in the first day's battle at Cambrai, and reached its objectives, doing good work at Pam Pam Farm and on the road leading to Masnières. On the 27th it went into battle at the village of Fontaine-Notre-Dame. The fighting in the streets was deadly. The upper windows of the houses were full of German machine-gunners, and the 6-pounder guns of the tank did great execution among them. All its crew were wounded, some severely, and the commander was struck in the eye. Notwithstanding, they brought the tank back to the rallying-point.

After this it was commanded by 2nd Lieut. J. Munro, and was re-numbered 6/36. It was with its battalion, with the 6th Corps, during the great Carman attack in the spring of 1918, and took part in all the fighting while we were on the defensive. It was finally handed in to workshops on June 19th, 1918.

The suggestion has already been made that it should be presented to the Malay States as a memento of the Great War, and of the generosity of Mr. Eu Tong Sen.

In April 1917, at the time of the Battle of Arras, I was invited to tank headquarters for a conference, and my diary of that period may be of interest:—

April 13th.
I was met by General Elles, Major Fuller, Lieut.-Colonel Uzielli, Lieut. Foote, Major Butler, Lieut. Molesworth and Colonel Searle. There was great news of the battle. Tanks had taken Thelus (Vimy), the villages of Tilloy, Neuville Vitasse, Monchy

le Preux, Riencourty Herdicourt; two tanks were cut off and captured, owing to no infantry support.

The proposed programme for the stay was as follows:—
Workshop Camp.—Saturday.
Battle front north of Arras.—Sunday.
Ditto, south of Arras.—Monday.
Technical meeting with General Butler.—Tuesday. General Baker Carr's brigade had been into action.

The general impression was that the men and officers were magnificent.

At the start of the battle, some brigadiers did not want tanks for fear of drawing fire on their troops, but all begged for them afterwards.

Five machines got bogged going into action. Ditching still the trouble.

Must have a guard for the Lewis gun.

April 14th.

I arrived at Bermicourt on the evening of the 13th, together with Captain Saunderson, Lieut.-Commander Barry, Lieut. Shaw, Captain Symes, Major Wilson, and Lieut. Thorneycroft.

Major Fuller gave a lecture on "what the tanks had done, and illustrated his remarks on a blackboard. He mentioned that the greatest difficulty was the bringing up of supplies of petrol, water and ammunition to the tanks in action. Among other things they had to bring up 20,000 gallons of petrol, and to have four dumps and two lots of 500 men each carrying.

I visited headquarters workshops. There were six buildings, to be later increased to nine. There was only one workshop building, and all repairs were done in hangars, one large and two smaller.

In the afternoon I saw a machine fitted with wooden spuds, using a spare new engine for experimental purposes.

April 15th.

I motored to Arras with General Elles, Colonel Uzielli, Major Wilson, and Captain Saunderson. We changed into a light car, put on steel and gas helmets and went through the old German lines, and got out at Beaurains, which was totally destroyed. We walked up Telegraph Hill and through the Harp, and saw all the tank tracks, also one tank which had been hit by a shell.

The driver had his head taken off. No damage was done to the machine except to the front plate and petrol tank. We saw many dead Bosches lying about, also very deep dugouts. On reaching Tilloy, saw German shrapnel bursting between us and Monchy le Preux, also gas shells. Saw there a German *cupola* (the same as in Invalides, Paris) on a carriage, with shafts for three horses, which had been captured by the French.

We returned to Arras, where I found my old friend Sidney Charrington in command of C. Battalion. He said that the tanks had done magnificently.

We motored to Noyelle Vion, and there saw an old friend. General Haldane, Lord Loch and General Tulloch. They said that the tanks had done splendidly and had taken Tilloy, Monchy and Evianeourt. General Haldane said he would write to the Adjutant-General telling him what a success the tanks were.

Returning in the evening, Harry Dalmeny rang up and invited General Elles and myself to dine at General Allenby's *château* with General Bols, General Lecky, and General Sillem.

General Allenby said he had not believed in tanks before, but now thought it was the best method of fighting, and would not like to attack without them.

April 16th and 17th.

I talked with General Elles in the morning, and at about twelve o'clock left for headquarters of the 18th Corps, Hautville, with Captain Saundeerson, to see my brother, a major in the South Irish Horse. I found that he was at Pas, where I arrived at three o'clock. We had a very excellent lunch of bully beef and biscuits. I had an appointment with General Butler at 4.30, so had to leave in a hurry. We met a lot of cavalry on the return journey, by way of Doullens and St. Pol, and arrived at 5.15. General Butler did not turn up until 6.30, when a Conference was held with General Elles, Major Wilson, Captain Saunderson and Captain Symes.

April 18th.

A technical meeting was held in the workshops. Snowing hard at lunch time. Arrived in Paris at eight o'clock, after a four and three-quarter hour's trip *via* Hesdin.

Lunched with Captain Leisse, and at 3.30 met Captain Marais, A.D.C. to Colonel Girard.

At 6.30 handed drawings to M. Breton.

April 19th.
Went to Marly.

April 20th and 21st.
I lunched with Mr. Lloyd George at the Hôtel Crillon; Lord Esher, Sir Maurice Hankey and three other generals were also present. I met General Maurice, the D.M.O. at the War Office, after lunch.
I got the passes extended and started for St. Chamond. Spent Saturday at St. Chamond and saw our tanks, fitted with their petrol transmission, ascend a hill of 55°.

April 22nd.
Returned to Paris. Saw Colonel Girard.

April 23rd.
Major Wilson, Captain Saunderson and I were entertained at lunch at Quai d'Orsay by the Minister of Inventions, M. Breton, General Mouret, General Estienne, Colonel Challiat, etc., and in the evening were entertained by Colonel Romailho at Rest Viel to meet all the tank experts. General Estienne explained the first battle with French tanks, the Schneider tanks, at Juvencourt. They were not a great success, getting ditched very easily and being quickly set on fire by high explosive shells.

April 24th.
Motored with Major Wilson and Captain Saunderson to Amiens; arrived at 1.30 and lunched with General Butler; went at 2.45 to house in Rue Gloriette. General Wilson, Duncannon, Field Marshal Sir Douglas Haig, General Davidson, and Sassoon arrived. Shortly afterwards, Nivelle and his staff arrived. Haig and Nivelle had a long conference. Afterwards all had a talk for half an hour. Haig then asked me in and Butler joined us. Haig said that he would do anything to help me; that a division of tanks was worth ten divisions of infantry, and he probably underestimated it; told me to hurry up as many Tanks as I could, not to wait to perfect them, but to keep sending out imperfect ones as long as they came out in large quantities, especially up till August. He said the tanks, after aeroplanes, were the most important arm of the British Army, as they were such tremendous life-savers. He asked who at the War Office did not believe in them. I replied that the A.G.'s department recruited

my men. He agreed that it must be stopped. He then congratulated me and thanked me.

Left for Boulogne *en route* for England.

During this visit I had a discussion with the officers of the Tank Corps about driving, and said that the skill of the driver had much to do with the success of a tank, and that a good driver could get a great deal out of a tank which was not mechanically perfect.

I challenged the Tank Corps to a race between Squadron 20 which was testing the tanks for my department, and any crew chosen from the whole of the Tank Corps, in one of their tanks, over any course they liked to select. I said that I would back my fellows to beat the others for £100. The challenge was accepted but not the bet, and the following letter from General Elles explains the result:—

Dear Stern,

This by Weston, whose visit has been most useful. His crew knocked ours by one and a half minutes out of sixteen minutes; but give us a month, as stipulated, and I shall be seriously inclined to take your money.

Searle back yesterday; tells me that the help he got from your fellows was quite immense. Indeed, the results are amazing, and we are all very grateful.

Yours sincerely,
Hugh Elles.

To these hurried notes in my diary can now be added a more connected account of the part which the tanks played at Arras.

Sixty in all of Mark 1 and Mark 2 went into action. They did not take there the leading part which they took in later battles, by concentrated attacks on single objectives, but were distributed among the different corps for minor "mopping up" operations.

As early as January the tank reconnaissances, preparation of supply dumps, tankodromes and places of assembly had begun. At this time there were no supply tanks in existence, and all supplies had to go forward by hand. It was reckoned that each supply tank would have saved a carrying party of 300 or 400 men. In all these preparations there was only one serious mishap. The night before the attack a column of tanks was moving up from Achicourt by the valley of the Crinchon stream. The surface of the ground was hard, but underneath, in places, it was a morass. This was only discovered when six tanks broke through the crust into mud and water. After hours of labour they were

H. RICARDO ESQUIRE

Tank presented by Mr. Eu-Tong-Sen

got out, but too late to take part in the first attack.

Up to the day before the attack the weather had been fine, but in the early morning of that day came a storm of rain and snow. It was bad for the infantry but worse for the tanks. On the heavily "crumped" ground of the Vimy Ridge, now soaked with rain, they could not operate at all, and were withdrawn to the fighting further south. Here they were more successful, but played no very great part in the first day of the attack, partly because of the bad state of the ground, partly because of the rapidity of the infantry's advance.

It was on the third day, April 11th, that their chance came. That day three important tank attacks were made. The first was made against the village of Monchy-le-Preux, which was not only strongly fortified, but from its ridge dominated the surrounding country. It was a tactical point of extreme importance. Six tanks advanced against it. Three were put out of action on the way, but the other three reached the village and enabled the infantry to occupy it.

Monchy captured, the cavalry moved forward. By all accounts the Germans were now thoroughly demoralised, but so long as they had a few steady machine-gunners who stuck to their posts, any effective cavalry advance was impossible. Only the tanks could have cleared the way, and it was the lack of a tank reserve to fill the places of those put out of action which gave the Germans time to restore their defence.

The same day, on another part of the battlefields four tanks worked down the Hindenburg Line, penetrated into the country beyond, and for over eight hours, unsupported by infantry, fought the Germans wherever they found them, killing great numbers. In the end all four returned safely to their own lines.

The third tank attack of the day was the most interesting, although the operation as a whole did not succeed. It was delivered against Bullecourt, and for the first time tanks were given the principal part. Although the operation failed, and it was found impossible to hold what the tanks and infantry had captured, it was made clear that the tanks could do those things for which until then massed artillery had been thought to be indispensable. They could cut the wire and they could protect the infantry with a mobile barrage in place of the creeping barrage of the guns.

The results of the conference at tank headquarters with General Butler and General Elles, on April 16th and 17th, should also be given in more detail. We talked over the improvements suggested by the experience of the battle, and I said that whatever could be done to improve Mark 4 during the summer would be done, but that I had a design with me which I hoped would meet all demands. It was for a Mark 6 tank, an improved Mark 5. This drawing I left with them.

General Butler then laid down the three types of machines required:—

1. The *heavy tank,* a tank which, owing to its weight, most have a special railway truck to convey it.

2. The *medium tank,* which can travel on a standard train in France.

3. The *light tank,* which can travel on a lorry.

I told him that I already had a complete wooden model of the Mark 6 which would be ready by the 30th April, with a real Ricardo engine in it. This would embody all the improvements required, greater speed, lighter loading and more ease of control. I also put forward the "Whippet,'" which has already been explained.

It was arranged that a conference should be held in London on April 30th to decide on the programme needed at once—not in preparation for next year's requirements, but to keep continuity of progress in design and output. This conference was to decide on our future policy, which must be governed by the demands of the Flying Corps and motor traction, both of which necessarily took precedence of tanks.

I suggested that lorries might be got from America. Her lorries were second to none, whereas her big engines and flying engines were not sound enough for our special enterprises.

General Butler suggested that the establishment of 540 would be enough. I replied that we ought to try to make progress; unless we increased, we should go back. I thought we could probably produce 300 a month of the heavy tank, and that 1000 ought always to be kept in commission as war establishment.

General Butler then queried the capacity of the country for spare parts, but Major Wilson explained that we were taking measures to eliminate the great wear and tear both by better design and better material. I cited the commercial success of the motor-bus, which originally suffered from stupendous wear and tear.

The question of armour-plate was discussed with Captain Symes. General Butler suggested Uralite, but Captain Symes explained that the best protection was to have armour-plate of the proper thickness.

Finally, General Butler said that he would tell the adjutant general of the value of the tanks, and the necessity of allowing us to have the men we needed.

Mark 4 machine

Mark 5 machine

SCENE ON THE FLANDERS BATTLEFIELD

THE STATE OF THE GROUND IS SHOWN HERE, OWING TO THE
HEAVY RAINS. A TANK IS SEEN HALF UNDER WATER.

CHAPTER 7

Fighting the War Office
May 1917 to September 1917

On returning from France I wrote to General Butler:—
Firstly, I wish to say how splendidly I consider the tanks have been handled and manned in the last engagement—a veritable triumph of training and organisation. But this is not the object of this letter. It is to ask you if the commander-in-chief could write officially to the army council informing them of the value of the tank as an established unit of his armies. My reason is that, owing to the great secrecy adopted in the early days, it is still looked upon as a surprise experiment which, after the original surprise, has practically no future.
If the commander-in-chief could write a letter to the minister of war as head of the army council and ask him to transmit it to the Ministry of Munitions it would give the tanks in England the serious position which he would undoubtedly wish them to have, and lighten the burden of those toiling for you at home.

As a result of this letter, Sir Douglas Haig wrote a letter to Lord Derby, the secretary of state for war, in which he said that the importance of tanks was firmly established and that there should be a special department at the War Office to look after them.

On May 1st Lord Derby called a meeting at the War Office at which were present General Sir Robert Whigham, the deputy chief of the imperial general staff, General Sir Neville Macready, the adjutant-general, General Furse, the master-general of the ordnance, General Butler, Dr. Addison, the minister of munitions, Sir Eustace d'Eyncourt and myself.

After Sir Douglas Haig's letter had been read it was decided that a tank committee should be established, with two representatives of the Ministry of Munitions, the master-general of the ordnance (representing the War Office), a general (representing the heavy section of the Machine-Gun Corps) and a chairman, who should be a general, not under the rank of major-general, with experience of fighting at the front.

On May 4th I saw Lord Derby and suggested that the committee which had been proposed on May 1st would be quite unsuitable, and that we ought to have a committee of people who could devote their whole time to tanks. The essential requirements were tactics, personnel and material. Therefore one person should devote his entire time, with agents both at Wool and in France, to looking after tactics, another to personnel, while the Ministry of Munitions would look after the design and manufacture. This should be the nucleus of the committee, and other people could be added as the War Office thought fit.

Lord Derby was inclined to agree with me and said he would get something written out. He suggested that the parliamentary secretary would probably be a good man for a chairman. As soon as he had drafted something, he would send for me to discuss the matter again.

Although we had at this time, and were to have later, enormous difficulties with the War Office, we were still on the best of terms with them, as the following letter from General Furse will show.

My Dear Stern,
I tried to get you on the telephone this evening to tell you how delighted I was to hear that those extra plates had been pushed off to France today for the Mark 4s. That is really a fine performance especially in this filthy time of strikes. Bravo!

The committee was appointed, with General Capper as chairman. A month before he had never seen a tank.

At the first meeting I spoke as follows:—

Sir Douglas Haig told me on April 24th that the tank had been established as one of the most important arms of the British Expeditionary Force, that it was a great life-saver, and that battles could not be won without huge losses except with tanks.
This new arm is making very rapid progress. In 1916 we produced 150 tanks; in 1917 we shall produce 1500 tanks; in 1918 we can produce 6000 tanks.
I do not believe that the standard organisation of the War Of-

fice can be imposed on this development without great injury. The connection between the fighting tanks in France and the department responsible for design, supply and development, must be direct and instantaneous. What possible advantage can accrue from the passage of ideas, and requirements through numerous departments of the War Office?

I wish to suggest that the rapid development can be achieved by the appointment of a director-general of all military tanks in England and abroad, solely responsible for everything to do with tanks, except when detailed for action in the field. Major-General Capper has already been appointed general in command of tanks, but to my mind he requires these powers. His position as regards tanks should be similar to the position which Sir Eric Geddes held in England and abroad with regard to railways.

The director-general of military tanks should leave all questions of design, manufacture, supply, inspection, labour, materials and transport to the mechanical warfare supply department, and should communicate his requirements direct to the director-general of that department, and he should approve the design of tanks before manufacture.

The duties of the director-general of mechanical warfare supply should be:—

1. Research.
2. Design.
3. Supply.
4. Transport.
5. Storage.
6. Repair, including central workshops at home and abroad.
7. Spare parts.
8. Any special tank Stores.
9. Inspection.

By this organisation, the users and producers would be directly in touch, and full information would be available at any moment, without passing through unnecessary departments.

The fighting force could inspect and take over from the suppliers at the central workshops, at home and abroad, tanks ready for action.

The study of wear and tear, spare parts, salvage, repairs, rebuilding, at home and abroad, would be simplified by being in the hands of one organisation.

These were the proposals which I made to the committee, but this new organisation appeared to Sir Eustace d'Eyncourt and me to be so dangerous to the free development of mechanical warfare that on the 18th of May I wrote a memorandum to the prime minister and the minister of munitions (Dr. Addison). It was a long document, setting out all the cases in which the War Office had disregarded expert advice with regard to tanks, only to find that the experts were right.

After describing the new arrangement, the memorandum went on:—

> The military authorities, who have not grown up with this new development, and who do not know the reasons for the different advantages and disadvantages which go to make a tank, are disregarding more and more expert opinion.
>
> For instance, the original tanks, Mark 1, were fitted with the Hotchkiss machine-gun, and the employment of this gun was only decided on after very exhaustive investigation and consideration of the special circumstances of tank warfare.
>
> On the 23rd November, 1916, at a special meeting the military authorities decided that the Hotchkiss machine-gun be abandoned, and the Lewis gun be substituted. This decision was reached against the advice of the mechanical warfare supply department.
>
> At a series of meetings again held by the military and reported on by a committee on the 6th May, 1917, it was decided that the Lewis gun was useless in a tank, and that the Hotchkiss gun was the only gun that could be used. The result is that once more a change has to be made, but too late to make the alteration as it should be made, and this year's tanks will carry converted Lewis gun loopholes and Hotchkiss ammunition in racks and boxes provided for Lewis guns.
>
> Originally, it was considered by the mechanical warfare supply department that male and female tanks in equal numbers was the best arrangement. In the winter of 1916-17 the military experts decided on the proportion of one male to two females. Last week they again changed their ideas to three males to two females, and there are insufficient 6-pounder guns for this purpose.

In consultation with the expert advice of mechanical warfare, it was decided by the military authorities that 100 tanks in France and 100 tanks in England were necessary for training purposes. On this decision, tanks were manufactured with mild steel plates instead of armour-plates, without the improvements required for fighting. This decision was only arrived at after very full discussion and after the question of this large number of tanks necessary for training purposes had been very severely criticised by the prime minister and Sir Eustace d'Eyncourt, but the military authorities stated that it was absolutely necessary to have this number of training tanks. Shortly after this, the military authorities decided to use sixty of these training Tanks for fighting purposes.

The reason given for using these practice machines was that the output of machines was late. The military authorities, though crying out for an early delivery of tanks, have never given the mechanical warfare supply department the assistance they required. At the same moment that the director-general of the mechanical warfare supply department was called before the war cabinet on March 22nd to answer for the delay of one month in the delivery of tanks, the War Office had made an agreement with the Ministry of Munitions, dated 16th March, 1917, that the two munitions which the army required most urgently were aeroplanes and guns. . . . At a meeting with Sir Douglas Haig, the minister of munitions, General Whigham, General Davidson, and Sir Eustace d'Eyncourt, on November 23rd, 1916, I stated quite clearly that to produce the tanks according to my proposed programme, I should require about 2000 workmen.

Instead of my getting these men, after repeated, continual, and continuous demands, men are being taken from the works, and only 275 out of the 2000 have been supplied.

The military authorities, being aware of the commander-in-chief's demand, had not put this department in a protected category, and it is only after repeated demands to General Butler, and a visit to Sir Douglas Haig last month (April 24th) at Amiens, that the mechanical warfare supply department now appears on the same list with guns and aeroplanes, and is thus protected.

The withdrawal of these training tanks for fighting purposes has resulted in a lack of trained men, and great delay in training the men, on which the whole success of mechanical warfare depends, has resulted.

The lack of training has also been due to a shortage of essential spare parts; a schedule was laid down by the military authorities in january on a much reduced scale to that suggested by the mechanical warfare supply department, and two months later certain categories were multiplied by thirty.

I then quoted from my letter to the minister of munitions of March 12th, protesting against the premature employment of tanks.[1]

The memorandum went on:—

The military authorities again disregarded technical advice, at a meeting on spare parts at the Ministry of Munitions, in the minister's room, at which the master-general of the ordnance, Sir Eustace d'Eyncourt, General Anley, Colonel Searle and Mr. Percy Martin were present (April 3rd, 1917). At this meeting, which had been called as a result of complaints to the War Council, the military authorities laid down that they required the same number of spare parts for training machines per 100 as for fighting machines per 100. I pointed out to them that the fighting machine was, compared with a training machine, a projectile, and that after going into action it would hardly require any spare parts; in other words, that training tanks would require spare parts, but fighting machines would require spare fighting machines. The technical side was overruled, and spare parts were ordered in quantities which are overtaxing the whole steel castings capacity of this country.

The technical advice that these machines were worth 200 *per cent*, or 300 *per cent*, more on dry ground than on wet ground was also disregarded. I explained that the wear necessitating the large number of spares was due to wet winter weather (namely, mud), and would be largely reduced by summer weather, and mudguards, which were being fitted. The advice was unheeded. A letter dated May 10th, from the heavy branch, France, states that all estimates for spares for fighting tanks, estimated on March 10th and confirmed at the meeting on 3rd April abovementioned, are to be halved, except the six steel castings which had been demanded in enormous quantities. These demands

---

1. See Chapter 6.

are reduced, in the case of four of them to one-sixth, and in the case of two of them to one-fourth. The total weight of the demand of April 3rd, covering 1600 machines over a period of eight months, was 10,500 tons, the amended estimate of May 10th for the same number of machines over the same period is 3400 tons.

The advice that tanks in quantity are essential to the success of an action in mechanical warfare has also been disregarded.

The preparations being made by the Ministry of Munitions should result in the shipment of between 40,000 and 50,000 tons of tanks during the fighting season this year. The military authorities have provided a single central workshop, one centre for the reception of this mighty host. This is against technical advice. The central workshop at the front, which we understood was to be used for running repairs, is now being used for the reconstruction of new types of machines such as 'supply tanks' and 'signalling tanks.'

General Anley was appointed general in command of the tanks in England, and the mechanical warfare supply department made arrangements for his offices to be next door to the mechanical warfare supply department. It was arranged that general demands from France for spares, tank stores, etc. from the Tank Corps should come direct through the general to the mechanical warfare supply department, and be shipped direct by the organisation of the mechanical warfare supply department, which ships tanks and tank stores, under control of its own officers and men, to the tank headquarters in France. Now, it is proposed to divert this through all sorts of channels in the War Office. Instances can be quoted by which the time taken in receiving a demand and fulfilling it under the new regime means endless delay.

The present suggestion to put mechanical warfare on the above basis at the War Office is, in my opinion, ill-considered and fatal to the success and progress of mechanical warfare.

It is proposed that two members of the Ministry of Munitions, a member of the War Office, and a member of the heavy section should meet under the presidency of a general not below the rank of a major-general. Major-General Capper has now been appointed administrative commander of all the tank forces, and is, I believe, on Sir Douglas Haig's staff, and also chairman of the tank committee.

I wish once more to put forward that, in my opinion, this committee, to be successful, must be put on a proper footing, and I consider it essential that the chairman, Major-General Capper, *must* be a member of the *army council*, and that the members of the committee should devote their entire time to mechanical warfare and be responsible to the chairman for the different branches which go to make up mechanical warfare, *viz*.:—

1. *Personnel*.

There should be a member responsible to the chairman for complete knowledge of requirements, and the study generally of personnel for tanks both in France and England, with a liaison with French tanks.

2. *Tactics*.

Another member should study the question of tactics in cooperation with the tank force in France and the tank force in England, and have complete information ready for the committee, with a liaison with French tanks.

3. *Supply and Design*.

All knowledge of supply and design should be concentrated in Lieut.-Colonel Stern and Sir Eustace d'Eyncourt, representing the Ministry of Munitions. The mechanical warfare supply department are in liaison with the French department on tanks design and supply.

The committee to consist of five members, three to be appointed by the army council, to devote their entire time to mechanical warfare; the chairman to be a member of the army council; the two members from the Ministry of Munitions, representing Supply and design, to be appointed by the minister of munitions, subject to the approval of the army council.

In this way, the chairman of the committee would have a permanent organisation with first-hand information on the personnel, tactics, manufacture and design of tanks, and, as a member of the army council, would be in a position to furnish the army council with the latest information on all these points whenever required.

This committee should be the recognised authority for [submitting recommendations as to the tactical employment of tanks and other mechanical warfare stores, add should advise on the numbers, training and equipment of crews, and on anti-tank expedients. The committee should be empowered to give final decisions (subject only to approval by the secretary of state for war and the minister of munitions) on all matters appertaining to design and equipment of tanks (whether for fighting, signalling, gun-carrying, supply or other purposes); to transport and storage of mechanical warfare stores, to repairs, including central and mobile workshops; to spare parts and any special stores, and to inspection.

The mechanical warfare supply department have hitherto carried out the experiments, design, construction, inspection, storage, and transportation. Their contract and finance sections have been under the director-general, subject to the supervision of the heads of these departments in the Ministry of Munitions, who have their representatives in this department. This system has proved entirely successful, and no faults have been found by the finance and contracts branches of the Ministry of Munitions except that it differs from the procedure in other departments, namely, that the local control of contract and finance departments, though supervised by the Ministry of Munitions, is directly under the director-general of the mechanical warfare supply department. The systems adopted by this department are used in the most up-to-date and successful businesses in the world, and have been approved by the different authorities of the Ministry of Munitions.

The question of transportation, storage and inspection, as above, should be left to the discretion of this committee to decide. My reasons for stating this are that mechanical warfare is in a process of great expansion and development, with violent and continual changes, and it would be fatal to try and impose on this new enterprise the hard and fast rules governing standard army stores. Of course, it is understood that the formation of this committee and the powers given to it are a temporary measure for this and next year's production.

If these recommendations are approved, I see every hope of carrying out the big programme which is on order this year, and the programme which has been foreshadowed for next year, but unless the new organisation is formed on these lines, with the flexibility suggested, I see no possibility of carrying out either this or next year's programme.

This document I signed, and Sir Eustace d'Eyncourt added his signature as concurring in everything that I had said.

In my covering letter to Dr. Addison, I explained that the memorandum did not concern supply, but concerned the whole question of mechanics as applied to warfare.

The prime minister arranged for a date for the discussion of this memorandum by the war cabinet. It had caused a sensation amongst the chiefs at the War Office, that any one should dare to question their ruling, especially a junior officer. Then General Capper came to see me. He told me that Sir William Robertson wished me to withdraw the memorandum, and said all the things which I had criticised would be attended to and altered. In consequence of this, although the memorandum had been circulated to all the ministers concerned, I withdrew it, with the result that nothing was done towards making the alterations which I had suggested.

I waited two months, and then, on July 23rd, 1917 I wrote again to the prime minister and to the minister of munitions, urging that the committee as set up was the wrong sort of body to control the construction of tanks.

Next day I wrote again to the prime minister on another and very important matter, the tactical use of tanks. The military authorities had agreed to make it a rule that they would not use tanks where the weather and ground conditions were very bad. In spite of this, in spite of the fact that the designers and builders had told them that over very heavily crumped, soft, muddy country tanks were practically useless, they had been using them in the mud at Passchendaele, and now it seemed likely that the army would cancel all orders for mechanical warfare.

In these circumstances I wrote:—

On January 25th, I had the honour of submitting to the Prime minister a memorandum for consideration (copy of which is enclosed) re mechanical warfare and the necessity of having a joint conference on tactics.

As a result of a discussion before the war cabinet between the general staff, Sir Eustace d'Eyacourt and myself, a meeting was held at the War Office on March 4th.

My suggestions in this memorandum were:—

    1. tanks must not be used over heavily crumped areas.
    2. tanks should be used in large quantities.

3. tanks should support tanks.

Since that date, we have fought battles at which tanks have assisted at Arras, Messines and Ypres.

I now propose to visualise the development of tactics, both enemy and ally, since August 1914, generally, and especially in relation to tanks and mechanical warfare.

1st Stage.

Swift shock-attack and counter-attack. German invasion of Belgium and France, ending in their defeat at the Battle of the Marne. Here is the end of mobile warfare.

2nd Stage.

TRENCH WARFARE

Defensive.—A long single line defended by wire, machine-guns and quick-firing field-guns.

Offensive.—Mass attack which failed—Neuve Chapelle and Loos.

At this point the thinkers (both German and Allied) saw that some new method of war must be adopted to break defences which defied any hitherto known weapons.

The Germans developed gas, with the idea of overcoming these defences by asphyxiation.

The Allies developed an armoured vehicle immune from wire and machine-gun fire, and capable of crossing trench defences. Both have failed to win decisive battles for the same reason, owing to lack of the big idea, lack of patience to wait until a new engine of war and its tactics could be fully prepared, organised and tested.

3rd Stage.

The Battle of the Somme in September.

Offensive.—The tactics of blasting the front lines of the enemy by means of masses of guns and ammunition, were adopted.

The advantages claimed were the protection of our infantry advance and the killing of the enemy in order to break through the angle line.

Defensive.—The Germans immediately began to appreciate our new offensive tactics and, by adopting numerous lines of trenches, gradually improved their tactics as an antidote.

They hold one or two lines very lightly, which are entirely obliterated by our massed fire. This fire makes it practically impos-

GUN-CARRYING TANK

SOME OF THE MANY STABLES
WHERE OUR TANKS ARE HOUSED

CENTRAL WORKSHOPS, TANK CORPS.
TANKS READY FOR ISSUE.

sible for our infantry, tanks or guns to operate. The result is that the enemy is safe from further attack until we have repaired the damage, but their guns, which are now placed further back out of our counter-battery range, create havoc among our advance troops who hold the newly-taken trenches unprotected.

The change in German tactics is clearly demonstrated by the number of guns captured by us:—

    200 at the Battle of Arras;
    40   " "   "   " Messines;

and a small number at the Battle of Ypres.

I think from the foregoing it is dear that to win a decisive battle, a gradual development from one form of attack by increasing, or perhaps decreasing, slightly the power of one arm of the force cannot produce the desired result. These changes can easily be countered by the defence. A total change of tactics is necessary—a surprise which the defence cannot deal with.

We still have that power of surprise in our hands.

Accumulate tanks and continue to do so until you have thousands, well-trained and well-organised tactically into an efficient, self-contained mechanical army. Continue to use a few at the front, if it is considered necessary to hoodwink the German command.

Press propaganda might also deceive the Germans in proclaiming the failure of mechanical warfare.

Make this effort a great enterprise of its own. Several types of tanks will have to be incorporated as the different new designs are introduced.

Finally this great force would consist of brigades of tanks of different designs, each organised for its own particular rôle; all organised under one head, who would be responsible for a mechanical army (trained with its complement of artillery and infantry, etc., until ready to complete its task) to win a decisive battle.

All arrangements for organisation, transport, etc., to be under this single head.

Here is a letter from a very important staff officer to whom I sent this memorandum:—

It is all a question of thinking big and thinking ahead and then behaving with sanity, holding fast to the principles of war and selecting your theatre of operations according to your weapon, or constructing your weapon according to your theatre. There are two theatres to consider: Flanders and Cambresis. The first has mud and the second wide and deep trenches. As I cannot imagine anyone choosing Flanders again, our difficulties may be spanning power. This requires very careful consideration.

The mud here is beyond description. I have never seen anything like it either on the Somme or the Ancre. The worst is our guns have destroyed the drainage system in the valleys, and little streams are now extensive swamps. Under our fire Belgium is going back to its primeval condition of a water-logged bog.

We can win the war if we want to, we won't if we continue as at present—so at least I think.

On July 27th Sir Eustace d'Eyncourt and I ceased to attend the meetings of the tank committee. We found that the three military members, who a month before had never even seen a tank, laid down all rulings even with regard to design and production. They were in the majority, and we could do nothing. That day I wrote to the prime minister:—

> A crisis has arisen in the relations between the War Office and the Ministry of Munitions *re* the progress and development of mechanical warfare, namely, design and production. . . . I continually pressed the War Office that they should establish a special department of tanks at the War Office, with whom my department could deal as a link between the fighting and production sides. The War Office has set its face against this most resolutely.
>
> I have had to visit nearly every department of the War Office on all sorts of vital questions and, naturally, with most unsatisfactory results. . . . The committee is now interfering in design and production, which, if allowed to continue, will result in chaos and disaster. I refuse to allow this.
>
> To put the matter on a proper basis is a most simple matter. Those conversant with the whole subject should be consulted, and the empty prejudices of the War Office cleared away.
>
> The proposition must be clearly stated and an organisation formed to suit the case, not some old dugout organisation which suits no modem requirements at all.
>
> There are two portions of this development—one fighting and

training, and the other design and production. For the former, a director-general of tanks, a member of the army council with a department at the War Office is essential. This department must have 'G,' 'Q,' and 'A' Branches as in the Royal Flying Corps.

We already have a director-general of mechanical warfare for the second portion.

As design and tactics are so nearly allied in mechanical warfare, there should be a connecting link in the form of a purely advisory committee, consisting of (War Office) director-general of tanks and deputy chief of general staff, (Ministry of Munitions) director-general of mechanical warfare, chief technical adviser, with a neutral chairman to give advice on questions of general policy.

We are ahead of any army in the world in mechanical warfare. This is due, after the initial decision of Mr. Churchill, to the pioneers of the work, namely, the mechanical warfare department of the Ministry of Munitions and the assistance of the Admiralty.

All the experimental work and testing, etc., is done by naval men lent by the Admiralty. *The War Office refuse all assistance.*

For the progress of a great technical enterprise of this sort the experimental and technical branches are all important and need a number of technical officers—younger men, who have the technical training.

The War Office refuse all these demands, refuse proper rank to those officers doing good work for this department, making their positions quite impossible. On the one side the War Office ask for complete co-operation with the front, *re* design, etc., and then make it practically impossible for me to send technical men over by refusing commissions and suitable rank to those who hold commissions.

All the technical men, consequently, go into the Flying Corps, where their cases are treated with consideration and common sense. If it was not for the Admiralty lending my department some 400 to 500 men for experimental and testing work, and giving my technical officers honorary commissions, I doubt if the tanks would ever have been produced in large numbers.

Sir Douglas Haig thanked me at the Battle of the Somme, September 17th, 1916, for the part we had played in producing the tanks, which were responsible for the greatest victory they had

had since the Battle of the Marne, and mentioned in dispatches Captain Holden, the deputy director-general of the mechanical warfare supply department, Major W. G. Wilson, Director, Captain K. Symes, director, and Lieut. W. Rendle, assistant director. These officers all hold high positions in my department. I have requested time after time that they should be promoted to the rank consistent with their duties. They all joined up for the war, at or soon after the outbreak. They are continually going to the front, but the War Office refuses, and General Capper refuses them all promotion.

When I had written this, I had an interview with the D.C.I.G.S., Sir Robert Whigham, who told me that it was intended to form a tank department at the War Office on the lines I had suggested.

I added a note to my memorandum saying that I had heard this was to be done, and urging that if it were done we should need also an advisory committee to guide the War Office on all questions of main requirements and other important questions such as tactics and training in relation to design. This committee, I suggested, should consist of:—

D.C.I.G.S., Major-General Sir Robert Whigham.
D.G., T.C., Major-General Sir J. E. Capper.
D.G., M.W.D., Lieut.-Colonel A. G. Stern.

Chief Technical Adviser, Sir E. H. T. d'Eyncourt.

It should be clearly laid down that the War Office give the Ministry of Munitions their requirements of the main programme on the advice of the committee as to type, but that the Ministry of Munitions must be entirely responsible for design, production, supply and transport, as hitherto.

The director-general of tanks should deal with the mechanical warfare department, through his "Q" Branch, for all details of design, manufacture, and supply of tanks, both armament and equipment.

The technical staff of this tank department at the War Office should be included in the "Q" Branch (T. 3).

Every effort should be made to discover and adopt methods of complete liaison with the fighting and training forces. It is suggested that a certain number of officers from these centres be attached for seven or fourteen days at a time continuously throughout the year to the mechanical warfare experimental grounds, and every facility should be given to the mechanical warfare department officers to visit the front and home camps as at the present time.

The War Office should recognise that the mechanical warfare department is exceptionally and intimately related in these stages of development with the fighting force and assumes, in certain of its departments, an absolutely military character, and that, subject to the usual approval, commissions and promotions should be granted, where necessary, in the Tank Corps.

While we were fighting for the proper management, and even for the very existence of tanks, at home, the following special order appeared in France, showing what they were capable of doing:—

The following Summary of the action of Tank F.41 (Fray Bentos), on the 22nd to 24A August, is published for the information of all ranks:—

F.41, accompanied by the section commander, crossed our lines on the 22nd at 4.45 a.m., near Spree Farm and proceeded eastwards with the objectives Somme Farm—Gallipoli—Martha House. The tank engaged and cleared Somme Farm. When in action against Gallipoli the tank commander was wounded, and while the section commander was taking his place, the tank became ditched. The tank continued in action, and our infantry entered Gallipoli and went forward to the north of it. At this time a hostile counter-attack developed, driving our troops back and regaining Gallipoli. Within sight of the tank the counter-attack was dealt with and broken by 6-pounder and Lewis-gun fire; heavy casualties were inflicted. F. 41 was now isolated. As our infantry were firing on the tank from the rear, thinking it had been captured by the enemy, the sergeant succeeded in getting back to them and informed them that we still occupied the tank and would hold out.

During the 22nd, the night 22nd-23rd, the 23rd, the night 23rd-24th, and the day of the 24th, the officers and crew, though all wounded, maintained their position, although heavily sniped by day and attacked each night. During these attacks the enemy actually got on top of the tank and brought a machine-gun up at very close range without effect.

At 9 p.m. on the evening of the 24th, the crew having maintained themselves for sixty-two hours, 500 yards in advance of and out of touch with our line, and having been seventy-two hours in the tank, decided to evacuate. This was successfully done.

Casualties.—Officers, 2 wounded; other ranks, 1 killed 4 wounded.

At this time also the gun-carrier tank went into action with great success.

Mr. Churchill had now become minister of munitions, but things were still unsettled. There was no final authority on the army side. There was a general at the front and a general in London, and a general at the training camp at Wool, and no one had authority in all three places. So we went on for some weeks, and on the 19th September I wrote to Mr. Churchill:—

> Lack of action and lack of decision will most assuredly ruin the chances of mechanical warfare for 1918.
> I have discussed the possibilities of mechanical warfare with the general staff, and I think they agree with me that mechanical warfare on a small scale is waste of treasure and effort.
> Every day and every hour that passes renders mechanical warfare on a large scale for 1918 more improbable.
> On July 23rd, you informed me that after a preliminary study generally of the ministry you would investigate the position of mechanical warfare. It is now two months since that date. Immediate action must be taken.

I then repeated my suggestions for an advisory committee, and went on to describe how the existing committee worked:—

> It has five members; the two members of the Ministry of Munitions have the experience and knowledge of developing mechanical warfare since February 1915, and have been responsible for the production of some 30,000 tons of tanks within a year, with hardly a complaint considering that it is an absolutely new development.
> This committee has three members who had never been in a tank till a few months ago when this committee was appointed.
> Every detail goes through this committee, it instructs our best drawing-office to be filled up with priority of a design which we, the experts, do not approve, with resultant delay to real progress; it fails to take the necessary action and risks which a head of enterprise must take at all hours of the day.

SUPPLY TANK

Offensive on the Cambrai front. A landship bringing in its prize; a 5.9 German naval Gun.

Since its existence, it has done nothing to further mechanical warfare, one of the most technical of all war developments. This country is pre-eminent in mechanical warfare, by reason of Mr. Lloyd George and Mr. Montagu deputising to the experts and allowing no red tape or out-of-date formulas to clog the wheels of progress.

The Ministry of Munitions' representatives Sir Eustace d'Eyncourt and myself, as all our protests were unheeded, ceased to attend the committee, but, at the urgent request of Sir Arthur Duckham, are again attending.

Complete chaos is the result of this ill-advised and ill-considered enterprise, the tank committee. I shudder to think of the harvest which you will reap next year if this is allowed to continue.

On September 21st I wrote a further memorandum to Mr. Churchill:—

At the request of Sir William Robertson, the C.I.G.S., I explained the latest tank developments and showed him the new one-man transmission at our experimental ground.

After a lengthy discussion I gather that he agrees that the science of mechanical warfare has reached a point when mechanical cavalry in large quantities, in conjunction with other arms, have a better chance than any known weapon of winning a decisive battle.

We believe that we have the design of such a machine.

He agrees that all our resources for manufacture of tanks should be devoted to the production of fighting tanks.

I wish to again draw attention to my minute of 24/7/17.[2]

Here I suggested that a tank effort should be a great enterprise of its own, all organised under one head. Now I suggest a still greater effort: let a great general organise our effort in conjunction with the Americans and the French; my department can organise the production in conjunction with the Allied generals.

My department could give all the drawings, specifications and our experience, and foster the allied output.

England could probably produce some 2000 machines of this type by July 1st, America probably 4000, France perhaps 500. This would give the Allies an overwhelming power for victory

---

2. The minute on tank tactics. See earlier pages this chapter.

to which no antidote at present exists.

Secrecy is essential.

On the 29th of September, 1917, Mr. Churchill called a meeting of the imperial general staff of the War Office, the general staff in France and the commanders of the tanks at the front and in England.

General Butler desired an entirely new programme. He said that the number of tanks got out depended on the number of men available, and Sir Douglas Haig estimated that there would be only 18,500 men available. He asked also that Mark 5s should replace Mark 4s, and that as few as possible of Medium A (the Whippet) should be made. Some "supply" tanks would be required, but he did not say how many. It was decided to meet again on October 10th.

The priority at this time was aeroplanes first, guns and ammunition second, mechanical transport third. Locomotives fourth, and tanks fifth.

On the 4th October the following recommendations were put forward to Mr. Churchill by Sir Arthur Duckham, Ministry of Munitions, member of council for tanks:—

Tanks

The situation at present may be summarised as follows:—

Both the design and supply of tanks have been in the hands of the mechanical warfare supply department, and they were the only people who had any direct knowledge on the subject of tanks until supplies were in the hands of the army. Through the use of tanks in battle and also through their use over old battlefields for the purpose of training, the officers in France are acquiring an actual knowledge of the use and deficiencies of tanks greater than that possessed by the mechanical warfare supply department. Meanwhile, the M.W.S.D. has realised that the design of tank now being manufactured suffers from considerable disabilities, and they have made radical improvements in design to meet these. The War Office during training with tanks at Wool are also finding troubles and are taking steps to overcome them, quite apart from the M.W.S.D., or the armies in France. Thus there are three different bodies making trials and doing experimental work without any proper co-ordination.

It has been necessary to form in the War Office a Department under a Director-General to control the requirements for and use of tanks, and the supply and training of the personnel. An effort was made to control the general design of the tanks required by a committee of the War Office on which the M.W.S.D. was represented by Colonel Stern and Sir Eustace d'Eyncourt, but the Tank Corps in France was not represented at all. This arrangement has proved unworkable, as the committee not only considered the general specification but also endeavoured to control working designs. I have discussed matters with Generals Whigham, Capper and Elles, and also with Colonel Stern and Sir Eustace d'Eyncourt, and I have obtained a general agreement in the following scheme:—

That the existing committee be dissolved and a new one set up composed of:—

General Capper (chairman).
General Elles, Tank Corps, France, or his nominee from his staff.
Colonel Stern, D.G.M.W.S.D.
Sir Eustace d'Eyncourt.

This committee should meet fortnightly in France and England in turn and its duties should be:—

(a) To discuss the requirements and possibilities of supply of tanks and formulate programmes.
(6) To advise on what line experimental work shall take and where it shall be carried out.
(e) To discuss tactics as affecting design.
(d) To arrange for a close liaison between the users of the tanks and the producers.

Only general specification as affected by conditions at the front would be discussed and settled by the committee, actual design would be carried out entirely by the M.W.S.D.

Liaison would be obtained as follows:—

(a) By providing a representative of the M.W.S D. with an office, an assistant and a clerical staff at the head repair shops in France.
(b) By a representative of the Tank Corps in France and a representative of the tank department of the War Office, being members of the design committee of the M.W.S.D.

On the 8th October a meeting was held under the presidency of Mr. Churchill, with representatives of the War Office and Ministry of Munitions, when it was decided to adopt Sir Arthur Duckham's recommendations.

On October 14th General Foch sent a message to me that he wished to inspect the latest development of mechanical warfare. i took him down to our experimental ground at Dollis Hill and showed him Mark 5, which was the latest type of heavy tank, manoeuvring and crossing wide trenches. He also saw the Gun-carrier. He congratulated me on the wonderful improvements and said, "You must make quantities and quantities. We must fight mechanically. Men can no longer attack with a chance of success without armoured protection."

Instead, however, of orders being given for thousands of tanks, as I had hoped, Mr. Churchill told me that the requirements for the army for 1918 were to be 1350 fighting tanks. This I determined to fight with every means in my power, and I told Mr. Churchill so. I then had an interview with Sir William Robertson, chief of the imperial general staff, and told him that the proposed preparations for 1918 were wholly and entirely inadequate. Sir William Robertson replied that this seemed pretty straight. I replied that it was meant to be straight.

Sir William Robertson was extremely polite and shook hands with me when I left.

TANKS OUT OF ACTION

LIEUT-COLONEL J. A. DRAIN, U.S.R.

CHAPTER 8

The War Office Gets Its Way
OCTOBER 1917 TO NOVEMBER 1917

On the 11th of October I asked for an interview with Mr. Churchill in order to put my views before him, for he appeared to be taking the advice of the War Office and not of the pioneers of mechanical warfare. I told him that I had served three ministers of munitions, that I had had the confidence and support of all three (Mr. Lloyd George, Mr. Montagu and Dr. Addison), that as a result I had done efficient work, and that without his confidence I could not make a success of mechanical warfare. He replied that I had his confidence, but that the War Office wanted a change made. The War Office, he said, accused me of lumbering them up with useless tanks at the front and of wasting millions of the public money.

Here I asked him to go slowly, as I wished to take down this astounding statement. In the opinion of the War Office, he said, there had been a total failure in design, no progress had been made, all the money spent on tanks had been wasted, and the belief in mechanical warfare was now at such a low ebb that they proposed to give it up entirely. Mr. Churchill paid me flattering compliments, and said that the country would reward me suitably for my great services.

I told him that I had fought against the forces of reaction from the day when the order for 1000 tanks was cancelled by the army council without the knowledge of Mr. Lloyd George, although he was secretary of state for war, and as a result of my protest was reinstated the next day; that time after time we had saved the War Office from wasting millions of money and going entirely wrong, and that our advice had finally been taken in each case in correction of the War Office's original action; I challenged Mr. Churchill to produce a single case where I had done anything to prevent progress and a free play of ideas, and I gave him two examples of the way in which I had worked. The transmission in the first tanks was not very satisfactory. Immediately after the first tank battle on the Somme I had put in hand every possible design of transmission, that we might discover the best. Again, on March 7th, 1917, I proposed to Dr. Addison to take over Sir William Tritton and Messrs. Foster's factory solely for experimental work, but Dr. Addison was unable to agree to this, as the future of tanks was at that time too doubtful.

So our interview ended.

On October 16th I was told by Sir Arthur Duckham that three generals at the War Office had asked for my removal. In this connection I will quote a letter received on the same day from the G.O.C., tanks in France, giving his views on what the department was doing for him:—

> You deserve a bar to your C.M.G.; and, seriously, we are extremely obliged to all of you for your very prompt action.
> Now we are very anxiously depending on you to solve two main conundrums which confront us:—
>
> (a) A device to get the Mark 4 and Mark 5 Tanks over a wide trench, and
> (b) Some very simple dodge by which we may be able to put on the unditching gear from the inside of the tank.
>
> We are trying a hook which is placed on a ring fixed either to a spud or to a track link between two wooden spuds. This hook is put on by hand through the aperture in the manhole, which we have enlarged to the breadth of the machine for the purpose.
> The weather has been ghastly lately, and the battle conditions are very trying, with all this rain and wet.

The whole trouble with the War Office was that I had pressed for a large programme of tanks—at least 4000—for the fighting of 1918, but the committee, against which we had continually protested, with its War Office majority of generals who knew nothing of tanks, had overruled me. Now, at a time when the decisions of experts were absolutely necessary in preparation for 1918, and when it was clear to us that enormous quantities of tanks were needed, the War Office programme was for 1350 tanks. Mr. Churchill told me that he agreed with Sir Eustace d'Eyncourt and me that quantities of tanks were necessary for 1918, but as minister of munitions he could not argue with the generals at the War Office about their requirements; his business simply was to supply what they wanted. This appeared to us a crying shame. We knew the thousands of casualties that the tanks had already saved in the attacks on the German machine-gun positions.

Next day Sir E. d'Eyncourt and I asked for an interview with Mr. Churchill. He refused to see Sir E. d'Eyncourt, and told me that, with regret, he had decided to appoint a new man in my place, and therefore there was no object in discussing the situation. He added that he was in power, and therefore it was his responsibility, and that he had taken the advice of the council member, Sir Arthur Duckham. I told him that I would not resign, as I believed it to be against the public interest, but that he could dismiss me.

Next day I received the following letter from him:—

Ministry of Munitions
Whitehall Place, S.W.
October 16th, 1917.

Dear Colonel Stern,

As I told you in our conversation on Friday, I have decided, upon the advice of the member of council in whose group your department is, and after very careful consideration of all the circumstances, to make a change in the headship of the mechanical warfare supply department.

I propose, therefore, to appoint Vice-Admiral Sir Gordon Moore to succeed you, and this appointment will be announced in the next two or three days.

I shall be glad to hear from you without delay whether those other aspects of activity in connection with the development of tanks in France and America, on which Sir Arthur Duckham has spoken to you, commend themselves to you.

Meanwhile I must ask you to continue to discharge your duties until such time as you are relieved.

Yours very truly,
Winston S. Churchill.

I had an interview with Sir Arthur Duckham on the same day, and he told me that Mr. Churchill was unable to persuade the War Office to have a larger number of tanks, but that as he was a believer in mechanical warfare, it was his opinion that America should be persuaded to arm herself with the necessary number of tanks for next year's fighting.

He told me that Mr. Churchill considered it my duty, as the War Office did not wish to develop mechanical warfare on a large scale, to undertake its development among the Allies, and chiefly the Americans. At this time I also saw the prime minister, and said that I was willing to undertake any duties which the country might call upon me to perform. On October 25th Mr. Churchill wrote to me as follows:—

> I was very glad to hear from you yesterday that you are ready to undertake the new appointment of commissioner for mechanical warfare (overseas and Allies) department, which I am now in a position to offer you. I need scarcely say that I should not have offered you this new appointment if I had not full confidence in your ability to perform its duties satisfactorily. You should settle the questions which may arise in regard to your staff with Admiral Moore and Sir Arthur Duckham. They will, I am sure, have every desire to meet your wishes and requirements. But at the same time I rely upon you not to ask for more assistance than is absolutely necessary.
>
> Recommendations as to the status and emoluments of any officers upon your staff should be made through the member of council in whose group your department lies. It seems to me that your first duty will clearly be to get into touch with the American Army and discuss with General Pershing, or his officers, what steps we should take to assist them with the supply of tanks.
>
> Perhaps I may take the opportunity of your assumption of new duties and responsibilities to convey to you on behalf of the Ministry of Munitions a sincere appreciation of the services which you have rendered to the tanks in the earlier stages of their development. No one knows better than I the difficulties and antagonisms with which you had to contend or the personal force and determination with which you overcame them.
>
> The fact that at this period in the history of the tank development I have found it necessary to make a change in personnel is in no sense a disparagement of the work you have done successfully in the past, and I can only hope that you will continue to apply to your new duties the same qualities of energy and resource which have already proved so valuable, and not allow yourself to be discouraged by the changes which it has been thought necessary to make.

In spite of what you said to me yesterday[1] evening, I still propose to submit your name to the prime minister for inclusion in the forthcoming honours list. I think this would only be right and proper so far as you are concerned, and that it would be helpful to you in the new work which you are undertaking. Unless I hear from you to the contrary my recommendation will go forward.

I enclose a formal statement of your new appointment and its duties.

On October 29th I accepted the position. On the same day I warned Mr. Churchill once more that the progress of design and the output of the tanks would most surely suffer. In the meantime Admiral Sir A. G. H. W. Moore had been appointed the controller of the mechanical warfare department.

Up to the date of his appointment Admiral Moore had never even seen a tank.

---

1. Mr. Churchill had said the night before that he was putting my name in for an honour, and I had refused it.

CHAPTER 9

The Anglo-American Treaty
NOVEMBER 1917 TO JANUARY 1918

As soon as the United States had entered the war, in the spring of 1917, I had called on the American military attaché in London, Colonel Lassiter, and asked him to come and see our tanks at the experimental ground, which he did. At that time he was the sole military representative of the United States in London and was fully occupied with all sorts of questions of war equipment, so that he was unable to devote any of his time to the question of tanks.

Sir Eustace d'Eyncourt determined, therefore, to see whether we could not arouse the enthusiasm of the Americans by getting the navy and the ambassador to see the tanks. In June Mr. Page, Admiral Sims and Admiral Mayo, with some forty naval officers, came to the experimental ground at Dollis Hill. They were delighted with what they saw. Admiral Mayo thought that the tank was the very weapon for the marines, and Mr. Page told me that he would cable to President Wilson that he considered it a crime to attack machine-guns with human flesh when you could get armoured machines, and machines, too, which he would never have believed capable of performing the feats actually carried out that day before him.

As a result of this Colonel Lassiter once more interviewed me. He said that the tanks must be for the army, not the marines, and agreed to cable for a technical expert to be sent over from the United States. This technical expert was Major H. W. Alden, who arrived on October the 3rd.

Now, at the beginning of November, having fought in vain for the greatest possible development of mechanical warfare in this country, I took up my new post and set to work to see what could be done with our Allies.

Mr. Churchill had given me the following letter to General Pershing and a similar letter to M. Loucheur, Minister of Munitions of the French Government:—

My Dear General Pershing,

The bearer of this letter, Colonel Stern, has been appointed by me commissioner of mechanical warfare (overseas and Allies) department, and proposes to establish an office in Paris in connection with his appointment.

I have instructed Colonel Stern to call upon you, as I desire that he should work in the closest relations with the American authorities. All communications from the Allies on questions of design, supply and experiments in relation to tanks are in future to be dealt with in the first instance by Colonel Stern, and, if only for this reason, it is desirable that he should make himself known to you.

The immediate object of his visit is to discuss with the French and American authorities, as my representative, the general situation of tanks and how best to develop their production. In particular he proposes, with Monsieur Loucheur's concurrence, to study the possibility of finding a factory in France in which tanks of a less recent design than that now in use can be converted into improved types. The possibility of furthering the assembling and even the construction of tanks in France for the Allies will also be considered by him in consultation with the French administration.

If you would be so good as to accord to Colonel Stern an interview he would be able to explain to you at greater length the precise objects of his mission.

As you are no doubt aware, he is one of the pioneers of mechanical warfare, and until assuming his present appointment was responsible for the supply of tanks to the British Army.

Yours sincerely,
Winston S. Churchill.'

On November 11th I had an interview with General Pershing and his staff and laid before him our proposals. He was very much in favour of the project, and said that he would give a decision within a few days. On November 14th he wired his approval.

A fortnight later I was able to tell Mr. Churchill that I had discussed the question of co-operation with Major J. A. Drain and Major Alden, representing the U.S. Army, and with M. Loucheur, representing the French Government, and that I had the honour to make the following proposals:—

(1) That a partnership of the U.S.A. and Great Britain should be incorporated for the production of 1500 heavy tanks at the earliest possible date, to be erected in France.

(2) A number of these tanks should be supplied to France if she should require them, in order to further the higher purpose of Allied unity.

(3) It might be convenient for France to supply an erecting shop without depleting her other supplies, but it might be wiser in any case to build a new erecting shop.

(4) No insuperable difficulties can be seen for the joint supply of components; the 6-pounders, ammunition and armour by Great Britain; engines, transmission, forgings, chains, etc., by the U.S.A.

(5) The design will be founded on the British experience with the U.S.A. ideas and resources. It will eliminate most of the faults of the present heavy tank in h.p., loading, crossing power, namely, the fault of not getting there.

The following is a rough comparison of the two types:—

	Mark 4	Liberty Type
Power	100 h.p.	300 h.p.
Loading	——	25% lower than Mark 4
Crossing power	11 feet	14 feet
Weight	28 tons	30 tons

The Liberty type should have a considerable carrying capacity in addition to its fighting power.

(6) Major Alden will collaborate in making the working drawings of new design before Christmas in London; all facilities of English engineers, draughtsmen and drawing offices must be put at his disposal. The design to be agreed upon by U.SA. and Great Britain.

(7) Labour other than skilled might be met by Chinese. The French authorities see no local difficulties in accommodating such labour.

(8) It is hoped to work up to 300 tanks a month after April.

(9) This will be a limited enterprise, and therefore a very high specified priority should be given by the three governments concerned in raw materials, labour, factories and transport.

(10) The entire management must be in the hands of the British and American commissioners, jointly with the French commissioner, where it concerns France.

It is hardly necessary to point out the great advantages of international standardisation or unity of manufacture.
    1. In production.
    2. In repair and spares.
    3. In progress.
    4. In training and tactical use.

On November 23rd, having had no reply from Mr. Churchill, I saw the Prime Minister and explained to him that the Americans had agreed to build jointly with us 1500 Tanks. I said that ten days had passed without a decision, and that they were chafing at the delay.

Next day I was informed by Sir Arthur Duckham that Mr. Churchill cordially approved my scheme for co-operation with the Americans, and suggested that M. Loucheur, French Minister of Munitions, should also be a party to the scheme.

M. Loucheur repeated that he could not join with us, as France had neither men, machinery, nor material to spare.

On the 26th of November the Ministry of Munitions informed the War Office that in order that the Allied Armies might be provided with tanks in adequate numbers, there should be the greatest measure of co-operation between the Ministry of Munitions and the departments of the Allied Governments responsible for the production of tanks. With this end in view a new department, known as the mechanical warfare (overseas and Allies) department, had been formed, and Lieut.-Colonel A. G. Stern, C.M.G., had been appointed to be head of it, with the title of commissioner. The department would act in accordance with the rules and procedure prescribed by the Inter-Allied Council or other co-ordinating authority in the United States or elsewhere, and also of any officers appointed to supervise all the munitions departments in Paris. A statement of the commissioner's duties was also forwarded to the War Office, with a request that the necessary notification be made within the War Office and to G.H.Q., France.

LIEUT.-COLONEL J. C. F. FULLER, D.S.O.

Tank factory, Neuvy Pailloux, France

Before this had come the news of the battle of Cambrai. There for the first time the tanks had fought as we had always wished, across good ground, without a preliminary bombardment, and in large numbers—over 400. With their help General Byng had won what up to that time was the greatest victory of the war, the greatest in the territory gain and the prisoners captured, and the greatest in its economy both of lives and ammunition.

A tank attack on such a scale had meant enormous preparation. Five million rounds of small arms ammunition, 165,000 gallons of petrol and 55,000 lbs. of grease were a few of the things collected in advance at the tank dumps. Moreover, the tanks had for this attack to be fitted with a special device. The span of the Mark 4 was ten feet, and it was known that in many places the Hindenburg trenches were twelve feet wide. Great cylinders of brushwood were constructed 4 feet 6 inches in diameter and 10 feet long. Each of these gigantic fascines was made up of a number of ordinary fascines bound together with chains. These chains were drawn tight by two tanks pulling in opposite directions. The fascine was carried on the nose of the tank and could be released by a catch from inside, the idea being that when a Tank came to a broad trench it would fill it up by dropping in its fascine, and so cross over. In the battle, however, it was hardly found necessary to use them at all.

Besides the tanks working with the infantry, each brigade had twelve supply tanks or gun-carriers and three wireless signal tanks, while thirty-two were fitted with towing gear and grapnels to clear the wire along the line by which the cavalry were to advance. Four hundred and twenty-two tanks in all went into action. They deployed on a line about 1000 yards from the enemy's outpost trenches, and at ten minutes past six, in a thick ground mist on the morning of November 20th, they began to move forward. General Elles led the attack in the centre, flying on his tank the Tank Corps colours.[1] It was

---

1. The history of these colours should be recorded. About the middle of August, 1917, General Elles and Colonel Hardress-Lloyd went to Cassel. For some time past General Elles had thought that the Tank Corps should nave colours like the Flying Corps. There at Cassel in a little shop the Colours were chosen. The colour scheme was to symbolise three things—mud, fire or the fighting spirit, and green field or "good going"; for the whole ambition of the Tank Corps was to fight its way through the mud to the green fields beyond. It is interesting, also, to note that the first Allied troops to enter Cologne were the 17th Tank Armoured Car Battalion, the leading machine of which was flying the Tank Corps colours. On reaching the Rhine, the flag was run up over the river.

the first British flag to fly in the Hindenburg Line. Close behind the tanks came the infantry. For ten minutes they advanced through the mist in complete silence. Then a thousand guns behind the British lines simultaneously opened fire, and their barrage of shrapnel, high explosive and smoke shells crashed down 200 yards before the advancing tanks.

The success of the attack was complete. The enemy ran for it or surrendered with very little fighting. Only at the tactical points did they make any serious resistance. In Lateau Wood there was a duel between a tank and a 5.9 howitzer. The gun turned on the tank, and with its first shell tore off most of the right-hand sponson, but none of the vitals were touched. Before the gunners could reload the tank was on the top of them and had crushed the gun down into the surrounding brushwood.

Other tanks, meanwhile, had topped the ridge and were hurrying down into the village of Masnières. Here was a bridge over the canal. It was the way to the next ridge and it still stood. A tank made for it, but as it neared the middle, the bridge bent and broke and the tank was flung into the canal. Other tanks came up and with their fire covered the crossing of the infantry.

Into Marcoing the tanks came so quickly that they shot down the engineers just as they were connecting up the electric batteries to the demolition charges on the main bridge. Everywhere the hurried retreat of the Germans could be traced by the equipment that they had thrown off as they ran.

While the attacking tanks were driving the German infantry before them, the supply tanks had moved up to their rendezvous; the wire-pullers had cleared three broad tracks of all wire so that the cavalry could move forward, and the wireless signal tanks had reached their positions. Ten minutes after the infantry had entered Marcoing, a signal tank had sent back the news that the village was captured.

In one place, the ridge by the village of Flesquières, the tanks unfortunately got too far ahead of the infantry. On the crest they came under heavy artillery fire at short range and suffered heavy casualties. Had the infantry been close behind them this loss would have mattered less, but as soon as the tanks were knocked out the German machine-guns came up again among the ruins. Flesquières was not taken until next day. Elsewhere, tanks and infantry worked in close co-operation and with complete success.

By four in the afternoon the battle was won, and, so far as the tanks were concerned, was tactically finished. There were no reserves, and all that could be done was to rally the weary crews, select the fittest tanks and patch up composite companies to continue the attack next day. But next day and the succeeding days of attack, although notable things were done, and on the 23rd Bourlon Wood was brilliantly taken by the 40th Division working with thirty-four tanks of the 1st Brigade, there was no longer the same close co-operation between the tanks and the infantry. New infantry had come up into the line.

That first day, however, had shown what could be done when tanks in numbers worked on a scientific scheme with the infantry. In twelve hours on that day, on a front of 13,000 yards, the attacking force had penetrated the enemy's lines to a depth of 10,000 yards and had taken 8000 prisoners and 100 guns. The prisoners alone were nearly twice the number of the casualties suffered by the attacking troops.

The number of the Tank Corps engaged was a little over 4000—no more, that is, than one strong Infantry Brigade, and that small body of men replaced the artillery, and made unnecessary the old preliminary bombardment. They did this also against trenches of peculiar strength sited on the reverse slopes of the main ridges, so that direct artillery observation of them was impossible, and protected by immensely thick bands and fields of wire. It would have taken several weeks of bombardment and many thousands of tons of ammunition to do what the tanks did in their stride, did without any warning to the enemy, and did more effectively.

Moreover, once the wire was broken and the infantry was at work, the tanks were able to work much more closely with them for their protection than had ever been found possible for the artillery.

On November 26th Sir Douglas Haig wired to the mechanical warfare department:—

> The tanks provided by your department have rendered very valuable services in battle near Cambrai. I beg you to accept and convey to all those under you whose skill and labour have produced the Tanks the grateful thanks of the army in France.

And in reply to a telegram of congratulations to General Elles, I received the following reply:—

Very many thanks. It was your battle too.

On November 26th Mr, Churchill sent for me. I spoke of the great success of the tanks at Cambrai, where they had been used—as they were meant to be used—in quantity and without any preliminary bombardment. I reminded him that not a month ago the War Office had accused me of lumbering up the front with useless tanks, these very 400.

I once more told him that the present organisation would never produce tanks in quantity or in time to win the war in 1918. Mr. Churchill asked what I meant and who it was I wished to place in charge of the mechanical warfare department. i replied that I had no wish to place anybody in charge, but simply to warn him that the present organisation would not produce tanks. This warning I repeated in a letter three days later when I wrote:—

> I hope you will allow me to express my conviction that the demands and preparations of the military authorities with regard to mechanical warfare for the fighting season of 1918 are entirely inadequate, and that changes which you have made in this department at this critical time (and which involve reconsideration of design and consequent loss of production) will most seriously affect even the efficiency of this programme for next season's fighting.
>
> This is also the considered opinion of my technical and commercial advisers.

On the 6th of December Mr. Churchill gave his approval to the Anglo-American scheme. I was appointed British commissioner, and Major James A. Drain of the United States Army was to be appointed American commissioner. The entire business was to be directly under the commissioners and to be called the "Anglo-American Commission."

Major Drain had been a general in the National Guard in the States, and as soon as his country declared war had come over to France, where he had been serving in the American ordnance department. He was a business man, and a man of great breadth of view. He saw at once that we should get the best results if the Americans adopted practically all our suggestions with regard to design, tested as they had been by our experience in the field, and set themselves to produce the intricate machinery required out of their vast resources.

The general type of design had been settled at a conference held at G.H.Q., France, on December 4th, at which were present representatives of all the fighting branches concerned. Details of design were to be in the hands of a committee under Major Drain and myself, consisting of Sir Eustace d'Eyncourt, representing the mechanical warfare department, Major H. W. Alden, representing the American Government, and Captain Green, representing the Tank Corps.

It had been agreed at the many meetings which had already been held that this programme, half the components for which were to come from England and half from America, should have priority after the War Office programme of 1850 tanks, and before any extended War Office programme.

In order that the necessary priority should be respected, both in England and in America, by the many different departments of the respective governments (experience having taught us that this priority must be over all departments), I invited the American commissioner to meet Mr. Lloyd George in order that he might suggest to him that a Treaty be drawn up between the highest authority in America and the highest authority in this country. This Mr. Lloyd George agreed to do, and the War Cabinet on January 8th, 1918, approved a Treaty which was signed later by Mr. Page, on behalf of the United States of America, and Mr. Balfour on behalf of Great Britain.

I give this Treaty in full. It is an historic document.

Agreement between the British and U.S. Governments for the Production of Tanks

The Government of the United States of America and the Government of His Britannic Majesty, being desirous of co-operating in the use of their respective resources for the production of the war machine known as tanks, and having considered the joint recommendation made to them by Lieut.-Colonel A. G. Stern, C.M.G., and Major J. A. Drain, U.S.R., whom they had appointed as their commissioners to investigate the possibilities of such joint production, the undersigned, duly authorised to that effect by their respective governments, have agreed upon the following articles:—

The above-mentioned commissioners are authorised by the respective governments—

(1) To build a factory in France, the cost of which and the running thereof is to be defrayed in equal parts by the contracting governments. The factory should be of sufficient capacity to produce 300 completed tanks per month and capable of being extended to produce at least 1200 tanks per month. The materials required for the construction of the factory shall be obtained in France and in England. The unskilled labour for the

LIBERTY TANK. HULL MADE IN ENGLAND, SHIPPED TO AMERICA AND FITTED WITH LIBERTY ENGINE.

A Tank going into action through our troops.

erection of the factory shall be supplied by the British Government. Skilled labour shall be supplied by the British or by the United States Government as the commissioners may arrange.

(2) To arrange for the production of, and to produce, 1600 tanks during the year 1918, or as many more as may be required and authorised by the respective governments, and to arrange for the provision of the components for these tanks in the United States and Great Britain substantially as follows:—

In the United States: engines complete, with starter and clutch, radiator, fan and piping, silencer, electric lighting, dynamo and battery, propeller shaft, complete transmission, including main gear-box, brakes, roller sprockets, gear shifting and brake control, track links and pins, rear track sprockets, hub and shafts, front idler hub and shafts, track roller, track spindles and bushings.

In Great Britain: bullet and bomb-proof plates, structural members, track shoes and rollers, guns, machine-guns and mountings, ammunition racks and ammunition.

(3) The respective governments undertake to give the necessary priority in respect at material, labour, shipping, and other requirements to enable the programme to be carried out in the most expeditious manner.

(4) It is understood that the tanks produced by the factory are to be allocated between the United States, France and Great Britain according to a determination to be reached later between the governments of the three countries, provided that the first 600 tanks produced shall be allocated to the United States Government, and *provided further that* the latter and the British Government shall each take one-half of the total number of tanks produced not sold to the French Government, unless unequal allocation between them shall be subsequently agreed upon.

(5) The price which shall be charged to the French, British and United States Governments, should there be an unequal allocation between the two latter, shall be £5000 sterling per tank, which price shall be subject to adjustment at the close of the operations occurring under this agreement, and the liquidation of all assets upon a basis of actual cost, such actual cost to include no charge for overhead by either government.

(6) The capital necessary to carry out this programme shall be supplied in equal parts by the British and United States Governments. Expenditure in France for labour and materials in connection with the building and running of the factory shall in the first instance be paid by the British Government. Materials purchased in Great Britain shall be paid for by the British Government, and those purchased in the United States of America shall be paid for by the United States Government.
An adjustment of the account shall be made every six months.

(7) It is further agreed that the United States Government shall replace the steel provided by the British Government for armour-plate. The replacement shall be in the form of ship plates and shall be made on or about the date of delivery of armour-plate to the factory, on the basis of ton per ton, the necessary allowance for difference in value to be made in the adjustment of the accounts.

In witness whereof the undersigned have signed the present agreement and have affixed thereto their seals.
Done in London in duplicate the 22nd day of January, 1919.

    (L.S.) Walter Hines Page.
    (L.S.) Arthur James Balfour.

Chapter 10

The Tanks Get Their Way
January 1918 to November 1918

I had succeeded now in increasing the probable supply of tanks from the very small number of 1350 ordered by the War Office by another 1500, but I was far from satisfied that we were making the progress that was necessary. I felt that if the Germans started making tanks they would probably overhaul us rapidly.

At this time I received the following letter. It showed what the men at the front thought of the tanks:—

I will first give you the opinion of one of my colonels. In three years fighting on this front, I've met no battalion commander to equal him in power of leaderships rapidity of decision in an emergency, and personal magnetism. I've met no man who would judge so justly what an infantry soldier *can* and *cannot* do.

He considers the tank *invaluable* if properly handled, either for the attack or in defence; but he realises, as I think we all do, that until Cambrai, the tactical knowledge shown in its employment was of the meanest order.

One other valuable opinion I've obtained. We have now with the battalion a subaltern, a man of about thirty—a very good soldier; a resolute, determined kind of fellow—who has seen a good deal of hard fighting. He commanded a platoon in our 11th Battalion in the big tank attack at Cambrai, and was in the first wave of the attack throughout. He tells me the tanks covering the advance of his battalion functioned under ideal weather and ground conditions, were handled with marked skill and enterprise in the capture of the first two objectives, covering an advance of about 3500 yards. The moral effect of the support given by the tanks on the attacking infantry is very *great*. He says his men felt the utmost confidence in the tanks and were prepared to follow them anywhere. The effect of the advancing line of tanks on the enemy infantry was extraordinary.

They made no attempt whatever to hold their trenches, and either bolted in mad panic or, abandoning their arms, rushed forward with hands uplifted to surrender. As long as the advance of the tanks continued—*i.e.* over the enemy trench system to a depth of from two or three miles—the total casualties incurred by our 11th Battalion (attacking in the first wave) were four killed and five wounded, all by shell-fire.

After the fall of the second objective, the advance ceased for some unexplained reason. They were told some hitch about Flesquières. The attack seemed to lose purpose and direction. tanks on the flanks began coming back. The battalion was ordered to attack five different objectives, and before the necessary plans could be communicated to subordinate commanders, orders were received cancelling the previous instructions. In a word, chaos prevailed. The afore-mentioned subaltern cannot speak too highly of the work of the tank commanders—nothing could exceed their daring and enterprise. He says he is absolutely convinced that infantry, unsupported by artillery, are absolutely powerless against tanks, and that no belt of wire can be built through which they cannot break an admirable passage for infantry.

Lastly, he makes no secret of the fact that it would demand the utmost exercise of his determination and resolution to stand fast and hold his ground in the face of an attack by enemy tanks, carried out on the same scale as ours. I may add that he is a big, upstanding fellow, a fine athlete, and afraid of nothing on two legs.

I give you his opinions at some length, because they are the *ipsissima verba* of a man qualified to speak from personal practical experience. Personally, I believe the tanks may yet play the biggest rôle in the war, if only the higher command does not damn them first by giving them the impossible to do, or, worse still, fail to employ them in situations where common-sense and past experience alike demand their use.

(Two days before the Hun attacked us at Bourlon Wood we lost

WHIPPET TANK

THE CANAL DU NORD. TANKS AND WOUNDED GOING THROUGH.

TANK AND PRISONERS WITH WOUNDED GOING THROUGH THE CUTTING IN THE CANAL DU NORD

three officers and some seventy gallant fellows trying to mop up a couple of enemy machine-gun nests— a bit of work a couple of tanks could have done *with certainty* without the loss of a man.)

In the situation described after the capture of the second objective, why should there not have been a responsible staff officer—G.S.O.I. say—right forward in a tank to size up the situation and seize opportunity, the very essence of which is rapid decision? In the early days of the war, forgetful of the lessons of South Africa, we put our senior officers in the forefront of the battle. Of late the pendulum has swung the other way. Surely the employment of a tank for the purpose outlined would enable us now to strike the happy mean.

In defence, as a mobile 'pill box' the possibilities of the tank are great—any man who has led infantry 'over the top' knows the demoralising and disorganising effect of the 'surprise packet' machine-gun nest. What more admirable type of nest can be devised? Continually changing position, hidden from enemy aircraft by smoke and dust of battle, offering no target for aimed artillery fire.

Half the casualties we suffer in heavy fighting after the initial attack come from the carrying parties winding slowly in and out through barrage fire, bringing up ammunition to the infantry, the Lewis and Vickers guns; all this could be done much more rapidly, surely, and with a minimum of loss, by tanks. For the future the tanks should relieve the artillery of all responsibility as regards wire-cutting. You *know* you can cross a belt of wire over which a tank has passed—you *hope* you can pass through a wire belt on which the artillery has played for a couple of days. As a business proposition, a Tank at £5000 will cut more wire in one journey, even assuming it does nothing else, than 2000 shells at £5 each, blazing away for a day—add the wear on the life of the gun.

In attack, one of the most difficult problems of the infantry is to get the Stokes guns far enough forward, with sufficient ammunition, to come into action against machine-guns or strong points holding up the advance unexpectedly. All this could be done by means of a tank with ease, whilst not only the small Stokes gun, with a range of 500-600 yards, can be brought forward, but I know of no reason why the '6-inch' Stokes, with a range of 1200-1600 yards, should not be brought forward by the same means, and be brought into action firing from the tank.

the tank has only one enemy to fear—the high-velocity tank-gun firing aimed shots from forward positions. I believe this danger can be minimised by means of escort aeroplanes attached during an action to every tank, and provided with smoke bombs to blind the gun position, if unable to silence the gun by machine-gun fire or by means of ordinary bombs heavily charged.

I have tried to outline some of the more obvious uses for which the tank is so admirably suitable. There is a well of this information yet untapped, not in staff offices, but in the minds of the platoon and the company commanders who have fought in the first waves of the attack with the tank, who have seen the difficulties it has to overcome and how it has met them or failed, and why.

Nothing has yet been produced in this war to equal the tank for doing by *machinery* what has hitherto been done by *men*. Nothing so well fitted to economise our man-power and reduce the appalling wastage which has hitherto characterised our efforts in attack, with gain instead of loss in efficiency.

We want thousands of tanks, both light and heavy, ranging from two miles to eight miles per hour, armed with machine-guns, armed with Stokes guns, unarmed and fast travelling for transport of gun teams to emergency tactical positions, and, lastly, a staff of trained minds to define the tactics of the tank—to refute criticisms based on ignorance, to collect, classify and investigate all available information and suggestions, so that, like an aeroplane, every new 'edition' of the tank is an improvement on the past.

I have written at some length, but the subject is big and attractive enough to be my excuse.

On the 8th of January I wrote to the prime minister as follows:—

We are entering the fifth year of the war.

I have watched the tactical changes of the armies on the Western Front. Tactics, born of necessity, such as trench warfare, the antidotes to trench warfare, and, again, the change in methods of defence against the novel massed artillery attacks.

I have watched with great care methods hitherto foreign to warfare used by both sides; these inventions failed to gain a decisive victory; they were used by both sides before they had been sufficiently developed either in efficiency, quantity, tactics or training.

I think that one should try to forget the past and imagine that our problem is the Western Front for 1918, oblivious of the past, except for its military tactical lessons.

It is dear to me that neither the Allies nor the Central Powers ought to be in a position to force a decisive battle on this restricted front during 1918.

Neither side has an overwhelming superiority in moral, men, guns or ammunition.

Both sides must try to establish a superiority in the line, and use that as a lever of advantage.

We should play up to the full, gamble to the full on any chance where we lead the enemy.

In aircraft we shall have no overwhelming superiority in design or scientific achievement.

We have this superiority in mechanical warfare. We have in our power weapons capable of killing unlimited numbers of the enemy, whereas our loss is limited.

A thousand tanks, with eight men in a tank and six guns, make a raid: the total wastage if all are lost is 8000 men. Sir Douglas Haig has estimated that a tank in attack has a value of from 300 men to a battalion. Here is an attack of from, say, 250,000 to 500,000 men carrying its own supplies, with no expenditure of shells except counter-battery work, with roads and railroads free from the extravagance of a non-mechanical battle. There is no limit, comparatively speaking, to the casualties such a force can inflict on the enemy.

I wish to suggest some arguments in favour of mechanical warfare for 1918 on the Western Front.

There should be four mechanical armies at different points on the line suitable for such warfare, all equipped and prepared to attack within a few hours in the form of a tank raid.

These raids to be an attack by surprise as at Cambrai, but with strict orders that a return should be made to our lines, obviating giving the enemy any opportunity of killing our infantry on equal terms, of attacking a salient or getting anything but a grave disadvantage in a counter-attack against our original positions.

My views are that in this way we can kill Germans and kill their *moral,* cause grove unrest and dissatisfaction between the men and their officers, between the people and the authorities, at their lack of efficiency in not adopting so formidable a weapon, used in overwhelming quantities by an enemy originally quite unprepared for war.

I have painted a very superficial picture of my ideas of the effect on the Germans.

Take the effect on our troops. Think of the infantry appreciating in its own delightful language the fact that it is about 100 to 1 on the tanks in these raids—really great tank battles, owing to the number of guns and men value of the tanks engaged, and that they can sit tight in their trenches awaiting the German counter-attack which must be carried out by infantry unprotected by armour against mechanical machine-guns.

The development of this great chance may strengthen the feeling in Germany, universal in the world, for a league of nations, by showing that the military caste is being beaten at its own game and losing its invincibility.

If this can be achieved by these means, before we have a possibility of overwhelming superiority in 1919, we shall have saved oceans of blood and mountains of misery.

If in your opinion my notions are in agreement with the reality of things, I wish to offer my services to study with the Allies concerned, and make a detailed plan, as far as possible, of such an organisation, complete, ready for operation.

This could, of course, only be achieved, with proper powers and with the complete good will and co-operation of the army and the War Office.

I heard later that the Germans were building large numbers of tanks. This drove me to make another great effort to get the War Office to order more. I communicated again with the prime minister, who said that if any group of ministers would support it, he would be ready to call a War Cabinet meeting in order that I could submit my views.

I had the pleasure of knowing Mr. William Brace, and asked him if he would bring the Labour ministers of the coalition to see the tanks and allow me to explain to them my views on the whole subject. I met Mr. Brace, Mr. George Barnes, Mr. Hodge, Mr. G. Roberts, Mr. Wardle, Mr. Walsh, Mr. Clynes, and Mr. Parker. After seeing the tanks and hearing what they had done— how they had saved thousands of casualties—they agreed to put forward to Mr. Lloyd George the suggestion that a War Cabinet meeting should be held with all concerned to press for the building of tanks to the full capacity of the country, subject to military advice and without interfering with the supply of guns and shells or with the requirements of the navy and merchant shipping.

On March 8th this War Cabinet meeting was held. General Sir Henry Wilson, the new chief of the imperial general staff, was present. He gave examples of the economy in men which had resulted from using the tanks. At Messines twelve divisions had been employed on a front of 16,500 yards, and after the first forty-eight hours' fighting our casualties amounted to 16,000 and the depth of our advance was 4000 yards. At Cambrai we employed only seven divisions, supported by tanks, on a front of 13,000 yards. Our casualties, after two days fighting, only amounted to 9500 men, and we gained in depth no less than 9000 yards, which meant approximately, with an equal force, half the number of casualties and double the gain in depth. Moreover, we saved at Cambrai a matter of 80,000 tons of ammunition.

The result of this meeting was that an extended tank programme of nearly 5000 tanks was adopted, and Mr. Churchill was asked to make the arrangements for it.

On April 8th Lord Milner, who up till this time had been cabinet minister at Versailles, and was now appointed secretary of state for war, came to see me at the offices of the mechanical warfare (overseas and Allies) department in Paris. I explained to him the development of mechanical warfare, and told him that the tanks had great power of destruction quite out of proportion to their own total cost of humanity, which was limited to eight men a tank. I told him that at the present time there was no central authority for the development of mechanical warfare, and that i considered it essential, for rapid development nationally, and internationally, that a special department, like the Air Ministry, should be formed, and that this ministry or board should be managed by those who had directed the development from the beginning, and were untrammelled by the vested interests of all the established branches of the War Office. In this way, a highly technical development could be carried out by a practical man with the advice of the military authorities.

I explained that I had been removed from my position as director-general of the mechanical warfare department on the demand of the War Office, because I had fought for the development of mechanical warfare, and told the War Office that their preparations for 1918 were entirely inadequate; that the programme had now been increased, too late, from 1350 to nearer 5000; that I had fought for the standardisation of mechanical warfare against continual change of design, and that standardisation was at last to be brought in by August 1918, again too late.

I said that we had fought our hardest to prevent inexperienced officers from ruining the one development in this country in which we had outstripped the Germans, but that instead of continuing its healthy growth under imaginative practical men, it had been placed under the heel of elderly service men, with the usual results; that the modem methods of standardisation and efficiency, untrammelled by army procedure and prejudice, had been stamped out; that the rules of the War Office made a civilian ineligible ever to become a soldier or to know anything about warfare, and that the Army Act was waved before the eyes of any junior officer who had ideas and dared to speak of them.

Finally, I begged him to see Sir Eustace d'Eyncourt, and to discuss the question of some proper authority to control and develop mechanical warfare.

From this date a new era of progress started for mechanical warfare at the War Office, with Sir Henry Wilson as chief of the imperial general staff and General Harrington as deputy chief. General Harrington believed in new methods and in mechanical warfare. He took the greatest trouble to give every assistance. About this time I had several interviews with him, and he told me that the Tank Corps was now to be brought into the army organisation, with the tactical side under the War Office branches concerned. Colonel Fuller was to be appointed to take charge of tanks in the department of the general staff.

I told him that, in my opinion, design, production and tactics were closely interlocked.

Tacticians could not make tactics without knowing quantities and types; producers and designers could not make quantities until at least a year after hearing the ideas of the tacticians; in fact, instead of the designs and ideas being thought out and criticised by the military tacticians, the tacticians, producers and designers should sit together and produce their plans together for consideration by the general staff.

General Harrington said that everything would be done to ensure the success of mechanical warfare, but that owing to many difficulties in the War Office it would take time, and at his request I promised that I would cease from forcing the pace until he had brought out his new scheme.

All this time I had been working in perfect accord with the French military and munition authorities.

On April 24th General Estienne's chief staff officer informed me that General Pétain had asked me to dine with him at his headquarters. I motored to Chantilly, and met him outside his villa one hour before dinner. He told me that he was a great believer in mechanical warfare, and asked me, if possible, to get powers from my government to form one central military school and training-ground for an Inter-Allied tank army at Châteauroux, with camps for British, French and American troops. He and General Foch were in complete agreement with the scheme, and their view was that tanks were infantry, and were absolutely essential in large numbers.

He asked me to see General Foch on the following day, but I had unfortunately to return to England to keep an important appointment. I very probably would have gone to see General Foch at once, but it was night-time, and no lights of any sort were allowed anywhere near Amiens, which at the time was extremely unhealthy.

On arrival in England the next morning I saw Sir Henry Wilson, who said he was going to France the next day and would propose the matter to Sir Douglas Haig.

Subsequently, although an Allied camp for an Inter-Allied army was not built, an Inter-Allied school for tactics was started at Bourron, south of Fontainebleau, where battalions of tanks of the different Allied nations were stationed for tactical instruction by senior officers under the presidency of General Estienne, commander of the French Tanks.

At a meeting at the War Office on June 25th, General Capper, who had been director-general of tanks and head of the tank committee, which had proved a failure, resigned; his post was abolished, and the question of a new authority to govern mechanical warfare was fully discussed.

Early in August, once more there was danger of a tank board being formed at the War Office, consisting of people who had no knowledge of tanks, and Sir Eustace d'Eyncourt, Admiral Sir A. G. H. W. Moore, controller of mechanical warfare department, Sir Percival Perry, his deputy and I, put forward to General Seely (who had become deputy minister of munitions) a scheme for making a new authority to deal with mechanical warfare. Just as at the War Office a new era of progress for mechanical warfare had started with the advent of General Harrington, so a new era of progress started at the Ministry of Munitions with General Seely.

Already we were beginning to see the results of the policy and the changes which the army council had forced on Mr. Churchill. Production had declined. It had fallen below the record of 200 tanks a month, which we had achieved in 1917, although the department now had much greater facilities for manufacture. It was, in fact, the one department of munitions of war which had not shown a continuous increase in output, and was producing only half of what it had promised.

Mr. Lloyd George, who, in spite of all his other activities and worries, continued his great interest in mechanical warfare, once again called a conference of the War Cabinet, as he was anxious about tanks, and wished to be assured that we should be able to achieve the increased programme which had been approved on March 9th after our meeting of the day before. At this conference it was decided to be absolutely necessary to have a strong board of competent men with the necessary authority to deal with questions both of design and supply.

The scheme for a tank board was put forward by Sir E. d'Eyncourt and myself as follows—

General Seely, as president, with Sir E d'Eyncourt as vice-president, Mr. Maclean (who had succeeded Admiral Moore as controller of production at the mechanical warfare department), Colonel Fuller, representing the general staff, General Furse, representing the army council, and myself, representing mechanical warfare (overseas and Allies). General Elles was added to the board, and the scheme was adopted. Later on Sir Percival Perry, representing mechanical traction, and Admiral Bacon of the munitions inventions department, were also added to the board, and, later still, General Swinton.

The new board proved a very great success. New ideas were received with enthusiasm; old-fashioned obstructions found no sympathy, and the programme for the year of 5000 English tanks had every chance of being completed. In addition, 20,000 light tractors, capable of carrying about five tons over *any country*, were ordered and in construction. These, though unarmoured, would make any army, both its men, its munitions, and its supplies, very mobile.

Before this, the Anglo-American commission had settled to work. An office and a drawing-office had been taken immediately in London, and an office in Paris. After some difficulty and with the help of the French Government, we had found a suitable site for our factory, and with it space for a training-ground. This was at Neuvy-Pailloux, some 200 miles south of Paris, and within easy reach of the two Franco-American ports of St. Nazaire, and Bordeaux. The whole of the material for building it, the equipment and the electrical power station were bought from England. The work was originally entrusted to the British Firm of Messrs. Holland and Hannen, Ltd., working under the direction of Sir John Hunter, director of factory construction for the Ministry of Munitions. In August, however, the work was handed over to Messrs. S. Pearson and Son, who completed the construction in November under the direction of Mr. F. J. Hopkinson. Here, Great Britain and the States were to build between them the 1500 super-tanks, each weighing forty tons. The whole of the armour, guns and machine-guns for these liberty tanks came from Great Britain, and the engines and internal parts from the States.

The first Liberty tank, however, was put together not in France, but in America. Major Holden, who had been my second-in-command, went to America to help in the new development of tanks there, and in July a hull, made in England, was sent over in order that the engines and gears which the Americans were to supply might be tested in it. This, the first super-tank, was completed by Captain L. R. Buckendale, of the United States Army, and Lieut. R. A. Robertson, R.N.V.R., who during 1916 and 1917 had superintended the whole of the inspection of the manufacture of tanks in England.

A great many experts had doubted if it were possible to use the Liberty Low Compression Flying Engine in a tank, but it came through the very searching trials with complete success. I had the following letter from Mr. Stettinius, the U.S.A. Deputy Minister for War:—

> War Department,
> United States of America,
> Paris,
> November 29th, 1918.

Edw. R. Stettinius, to Lieut.-Colonel Sir A. G. Stern,
2, Rue Edouard VII.
Paris.

My Dear Colonel Stern,

I have received the following cablegram from Washington,

which I believe will be of interest to you:—

Cable Number 72,

'Par. 1. With reference to your 584 par. 4 supplementing our 49 par. 2. Mark 8 field-tests in progress. So far no structural defects. Machine makes six miles per hour on high and has ample power for climbing. Has negotiated thirteen-foot trenches with its parapet repeatedly. Leverage of track brake foot-pedal had to be doubled to produce satisfactory steering. Original leverage would not lock one track under all conditions. No engine trouble experienced in actual tank tests. Reduction of width of reverse clutch slots and placing mufflers on top have been found necessary, as you suggested. Notify Stern, London, England. Goethals.'

Yours very sincerely,
Edw. R. Stettinius,
Special Representative.

Beside the 1500 tanks which she was building jointly with us in France, and which were to be distributed among the Allies according to the decision of the Versailles Council, America decided to build another 1500 at home, as well as many thousands of Renault tanks. She found, however, greater difficulty than she had expected in making the armour-plate and guns, so it was arranged that the second 1500 should also be assembled at our French factory on the same terms as the first, England supplying the armour-plate and guns, America the engines.

I believe that the joint working of the Americans and English in this way is unique in the history of the world. Many people had protested that we should never succeed in doing it amicably and successfully. All their fears were proved wrong. American and English officers and civilians! men and women, worked together in perfect accord. I know of no single instance of discord among us, and I know that the whole of my staff had never throughout the war found a finer *esprit de corps* than inspired this enterprise of the Anglo-American tank commission.

We were helped by the French Government in very many ways. It provided the greater part of the unskilled labour for building the factory, consisting of labour battalions of Annamites from Cochin China,

I will quote a letter from the French Ministry of War giving their

SMOKE BOMB EXPERIMENTS

TANKS GOING FORWARD TO CROSS THE HINENBURG LINE

views of the enterprise at Neuvy-Pailloux. It was written after the Armistice had been signed.

Republique Française
Ministère de la Guerre,
Direction de l'Artillerie Sous Direction
de l'Artillerie d'Assaut,
Paris, le 18 Novembre, 1918.

Le President du Conseil,
Ministère de la Guerre,
   to
Anglo-American Commission,
2, Rue Edouard VII,
Paris.

Confirming our telephonic message, we beg to state that the French armies will not need now any Liberty tanks. We wish, however, to receive two or three of these tanks fitted exactly as they would have been at the Neuvy-Pailloux Works. These would be used for experimental purposes.

We avail ourselves of this opportunity to declare emphatically that we have highly appreciated the efforts made by your commission in originating and pursuing the completion of a very extensive work, which would have greatly helped France in its struggle.

Very truly yours,
Aubertin,
Lieutenant-Colonel.

It will give some idea of the importance of the factory at Neuvy-Pailloux, when I say that the output of Liberty tanks (with a h.p. more than twice as great as the h.p. in any British tank up to that time) would have been as large, if not larger, in the first months of 1919, as the output of the factories in the whole of England. mechanical warfare and mechanical transport were now being developed on such a scale that beyond doubt they would have proved decisive had the war continued into 1919, but before they could be used on a great scale the war was over, and the Liberty tanks never went into action.

While we were preparing and building for 1919 through the summer and autumn of 1918, the tanks, in spite of the blunders which had limited and delayed the construction for that year, were playing a great part in the battles from Amiens to Mons.

In June and July, before the great offensive began, the tanks fought three actions. They were all three small affairs, but each was noteworthy. It had been unfortunate for the tanks that the great success of the first day's attack at Cambrai in November 1917, brilliant in its actual achievement and still more brilliant in its promise of what tanks could do, had been largely obscured by the unexpected and disheartening success of the German counter-attack. The three actions of June and July, small as they were, served to hearten those, and there were some, who had begun to wonder if there was indeed a future for the tank.

The first of these actions was a raid near Bucquoy, on June 22nd. It was carried out by five platoons of infantry and five female tanks, and was the first occasion on which tanks had attacked by night. It showed not only that they could manoeuvre by night, but that darkness was a great protection to them. The infantry was held up by a heavy barrage from trench mortars and machine-guns. Though reinforced, it could not advance, and the tanks went on and carried out the attack alone. One tank was attacked by a party of Germans, and its crew shot them down with revolvers. In spite of the heavy trench mortar fire not a Tank was damaged, and all returned.

On July 4th sixty tanks went into action with the 4th Australian Division against the Hamel spur, which runs from the plateau of Villers-Bretonneux to the Somme. It was an attack on a front of just over three miles, and was to be carried to a depth of a mile and a half. Not only were all the objectives reached, but each was reached by the time fixed in the plan of attack. The number of prisoners taken, 1500, was more than double the total casualties of the Australians, while only five of the tanks were hit, and the casualties of their crews were only sixteen wounded.

The co-operation between tanks and infantry came as near perfection as could be, and the Australians were finally convinced of the advantage of working with tanks. The full value of that conviction appeared in the greater battles that were to come.

In this attack Mark 5 tanks went into action for the first time, and more than justified all expectations of them. They all reached the starting-point in time. That, in itself, was an achievement. It showed the mechanical superiority of the Mark 5s over the earlier types. Their greater sureness and speed in manoeuvre were shown by the large number of German machine-guns that were crushed. Once a tank had passed over a machine-gun crew there was no fear that it would come to life again behind the attacking infantry.

Since the action was on a small scale, there had been no need to have an extended system of supply dumps. Each fighting tank carried with it ammunition and water for the infantry, and four supply tanks brought up the supplies for the engineers. The four brought up a load of 12,500 lbs. and had delivered it within 500 yards of the final objective within half an hour of its capture. Four tanks and the twenty-four men in them had done the work of a carrying party of 1250 men.

The same month for the first time British tanks went into action with French infantry. This was near Moreuil, some miles north of Montdidier.

Three French divisions attacked on a front of two miles. The tanks engaged were the 9th Battalion of the 3rd Brigade, seventy-five tanks in all, and they worked with the 3rd Division. After the battle they were inspected by General Debeney, commanding the First French Army, were thanked by him for the fine way in which they had fought, and as a sign of their comradeship in battle with the 3rd Division were presented with its divisional badge. From that day the men of the 9th Battalion have worn it on their left arms.

On July 15th the last big German attack was launched against Château Thierry and failed. It left the Germans holding a dangerous salient. Three days later Marshal Foch made his great counter-attack against the western flank of this salient, striking the first of those Allied blows which were to continue up and down the whole front without intermission, until four months later the German Army could fight no more. In this first victory the French Renault tanks played a conspicuous part.

Two days before the German attack was made the commander of the 4th British Army, which was holding the line before Amiens, was asked by G.H.Q. to submit a plan of attack. On August 8th, the attack was made on a front of ten miles. The attacking troops were the Canadian and Australian Corps, the 3rd Corps, three divisions of cavalry and the whole of the Tank Corps, except one brigade, which was still armed with Mark 4 machines, and was training its men on the Mark 5.

As at Cambrai, the tanks led the attack without any preliminary bombardment, but with an artillery barrage and a special noise barrage to cover their approach. The battle began at a quarter to five, when 430 tanks out of the 435 that had been assembled went forward.

Two of the brigades of fighting tanks were armed with Whippets, ninety in all, and worked with the cavalry, and besides the 430 fighting tanks there were numbers of others for supply and signalling.

The attack came on the German infantry as a complete surprise. The tanks appeared above it out of the morning mist, and the line, strongly held though it was, broke before them at once. It was noticed that the German machine-gunners, who had learnt already, in the smaller actions of June and July, that we had a new and faster tank, were much less tenacious than in any previous battle. They did not wait to be crushed beneath these great machines of thirty tons weight each, which came searching for them among the standing corn.

By the end of the day the attack had been pressed to a distance of over seven miles, but when in the evening the tanks rallied, it was found that 100 of them had been temporarily put out of action, while the crews of the rest were exhausted with the long distance covered and the August heat. Composite companies were hastily arranged, for there were few reserves.

Next day, when the attack began again, 145 tanks went into action. The total hit that day was thirty-nine, but in one part of the line, round Framerville, it was only one out of thirteen. This was because the infantry fought very skilfully to protect them. Infantry and tanks went forward together, and the riflemen picked off the enemy's gunners as soon as the tanks came under observation of the guns.

On the third day sixty-seven tanks were engaged and thirty were hit, and on this day the edge of the old Somme battlefield, pitted with shell craters, was reached.

On the fourth day there was no general attack, but a number of small operations against German strong points which still held out.

Within the next few days it was decided that the 3rd Army should take up the attack north of the Somme. The battle of Amiens was at an end. In the four days the 4th Army had gone forward from six to twelve miles on a front of over twelve miles, and it now held almost the same line that the French had held on July 1st, 1916. It had taken 22,000 prisoners and 400 guns. Of the part played by the tanks Sir Henry Rawlinson spoke in a special order of the day:

> The success of the operations of August 8th and succeeding days was largely due to the conspicuous part played by the 3rd, 4th and 5th Brigades, and I desire to place on record my sincere appreciation of the invaluable services rendered both by the Mark 5 and the Mark 5 Star and the Whippets . . . and of the splendid success that they achieved.

A GERMAN ANTI-TANK RIFLE

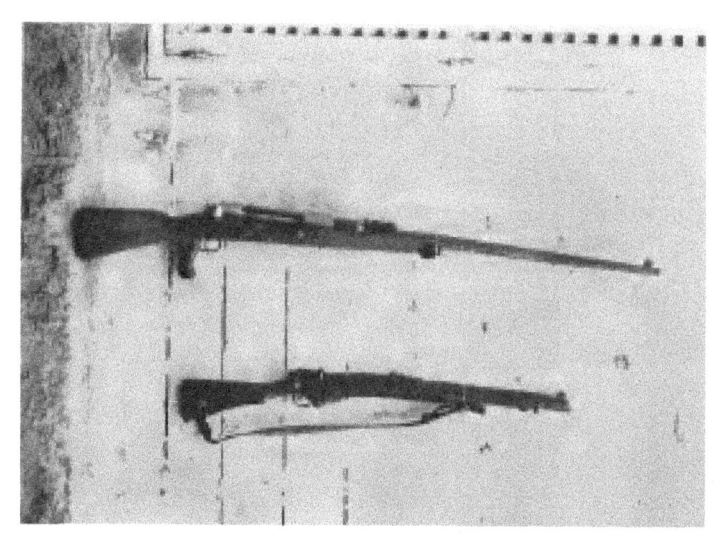

A GERMAN ANTI-TANK RIFLE COMPARED WITH A BRITISH

A German Tank

The battle had taught the Tank Corps some new lessons and confirmed the old. It had proved it to be a mistake to attach the Whippets to the cavalry. In the approach marches they could not keep pace with it, in the actual fighting they were kept back by it. By noon on the first day there was great confusion behind the enemy lines. The Whippets should then have been five or even ten miles ahead of the infantry, spreading the confusion and frustrating all attempts to restore order. As it was, they were kept far behind them by their orders to support the cavalry, for the cavalry, unable to take cover like the infantry, was compelled to retire to a flank or to the rear before machine-gun fire. The Whippets did some hard and gallant fighting, but co-operation between tanks and cavalry was proved to be to the help of neither.

Another important lesson was that while the tanks were the great aid and protection of infantry against machine-guns, they themselves, fighting as they now were, not across fortified positions, but over open country, needed the protection of the infantry against artillery fire.

These were valuable lessons, but the two crying needs were for still faster tanks, tanks that could have outstripped a retreating enemy and cut him off, and for a tank reserve. That lesson, which had been taught us first at Arras and then at Cambrai, but which the War Office had refused to learn, was repeated. The endurance of a heavy tank in action was three days. Without a general reserve, the real force of the tanks' blow was spent on the first day. They went on afterwards with tired crews, in battalions hurriedly rearranged and in much diminished numbers. Altogether in those four days of fighting 688 tanks went into action; 480 had to be handed over to salvage, and of the remaining 288, very few were fit for anything but a short attack, and all required a thorough overhaul. There was little time for it. The battle north of the Somme was to begin on August 21st. The enemy showed signs that he was preparing to retire between Arras and the Somme. He was to be attacked before he could do it.

The first attack started between Moyenville and Beaucourt-sur-Ancre. Once more it took the enemy by surprise. The tanks, when they crossed his first trenches, found candles still burning in them and a great litter of papers and equipment that he had thrown away. Moreover, he had adopted a new system of defence. His first line was very lightly held and his guns were withdrawn. The result was that few were overwhelmed and captured in the first surprise of the attack, and that in the second and third stages the tanks came under very heavy fire. To meet this new defence the older tanks were used against the first and second objectives. Then the new Mark 5s took up the attack, and then the Whippets. By the time the second objective was reached the mist that had hidden the first stages of the attack began to lift. Each Tank went forward, the centre of a zone of bullets and bursting shells. The fire was concentrated on them, and the infantry, in consequence, had few casualties.

Next day the 4th Army took up the attack as far southwards as the Somme. On August 23rd it spread south of the Somme. On August 26th northwards to Arras. The 1st, 3rd and 4th Armies were now attacking on a front of thirty miles, from Arras to Chaulnes. This battle was fought right across the battlefields of 1916 and 1917. It lasted a fortnight. It took from the Germans in captures alone 53,000 prisoners and 470 guns, and it culminated, on September 2nd, in the breaking of the famous Drocourt-Queant line, which in April 1917 we had failed to reach. It was an immensely strong line protected by great belts of wire. It fell to the tanks in a day. Except for heavy anti-tank rifle fire they met with little opposition—with far less, indeed, than had been expected, but one company of tanks alone destroyed over seventy German machine-guns. The gunners surrendered as the Tanks approached.

During the fortnight, except for one or two minor failures, every attack had succeeded, and succeeded, too, with casualties to the infantry on a much smaller scale than in previous attacks. The tank Corps had moved up to the battles of Bapaume and Arras straight from the battlefield of Amiens. They had had scarcely any time for repairs, for rehearsal with the infantry or for reconnaissance. They had been through a fortnight of almost continuous fighting.

On September 4th all tank brigades were withdrawn to G.H.Q. reserve to refit and reorganise.

Tanks were again in the line twelve days later and took part in the battle of Epehy. They fought that day under a heavy gas barrage, which forced their crews to wear their gas helmets for more than two hours on end. Then, on September 27th, began the third of the great battles, the battle for the Hindenburg defences, for that zone of entrenchments five to ten miles deep, heavily wired and drawing added strength, in front of Cambrai from the Canal du Nord, and between Cambrai and St. Quentin from the Canal St. Quentin. It was fought by the 1st, 3rd and 4th Armies over a front of thirty miles, from the Sensee River to St. Quentin, and it lasted fifteen days.

The battle was begun on September 27th by the 1st and 3rd Armies with an attack towards Bourlon Hill in front of Cambrai. Fifty-three tanks fought that day, some of them over nearly the same ground where they had fought in November 1917. The Canal du Nord was the chief obstacle before them, and between Marquion and Bourlon the Germans had so far trusted in it as to prepare no special anti-Tank defence. It was a great dry channel (for when the war came it was still unfinished), fifty feet broad at the bottom and twelve feet deep. Its banks were steep, but to make doubly sure the Germans, in places, had cut the bank into a vertical wall nine feet high. Yet all along the line the tanks crossed the canal, even climbing the nine foot wall. Bourlon Hill was captured, and next day the infantry were on the outskirts of Cambrai.

On September 29th the attack was taken up by the 4th Army further south. The American Corps fought in the centre with the Australian Corps. A British Corps was on either wing. The infantry were supported by 175 tanks. It was an attack on a large scale, carefully planned, its object being to cross the Canal St. Quentin and force the "Hindenburg" defences. In the northern half the plan miscarried. It miscarried because a preliminary attack the day before had only half succeeded. What should have been the British front line on the morning of the big attack was still in German hands. There was delay. The artillery barrage got far ahead of the infantry, and the Americans who led the attack suffered very heavy losses. The attack failed. Disaster also came on the 301st American Tank Battalion, which was working with an American division. It ran into an old British minefield of rows of buried trench mortar bombs. Ten machines were blown up; the whole bottom of several of them was torn away, and only two were able to support the infantry.

Further south the tanks of the 4th and 5th Brigades broke into the Hindenburg line; then the morning mists began to lift and it was found that, as a result of the failure to the north, the flank and rear of the attack were exposed. The later objectives had to be abandoned, but several tanks went into action on their own initiative without artillery or infantry support, and though they suffered heavily themselves, saved the infantry many casualties.

On the southern wing, though the attack was made in a dense fog, it was a complete success and the Canal was crossed.

Next day tanks were in action with the 1st Army north of Cambrai, where they used smoke clouds very successfully to hide themselves from the German gunners; and on October 1st, on the 3rd and on the 5th, they were again attacking with the 4th Army.

The second phase of the battle of Cambrai and St. Quentin began on October 8th, when the 3rd and 4th Armies attacked together on a front of eighteen miles between the two towns. Eighty-two tanks went into action that day, and at one place there was a duel between tanks. The Germans had captured from us one male and three female. With these they counter-attacked. The male was put out of action at once by a 6-pounder shell fired from another tank, and one of the females by a shell from a captured German field gun fired by a tank section commander. The other two females turned and fled when one of their own sex advanced to attack them. This was the second battle between tank and tank. The first had been fought in April with equal success.

Next day the attack was taken up again along the whole front of thirty miles; Cambrai was occupied, and by that evening, October 9th, the battle was over. The whole of the Hindenburg defences had been captured, and the attempt to hold them had cost the Germans, in captures alone, 600 guns and 50,000 men.

It was not only in Cambresis that the German trench system was broken. The same thing had been done in Flanders by the combined French, Belgian and British forces at the battles of Ypres and Courtrai, and now along the whole length of their line the British Armies followed their retreating enemy over open, unfortified country. But the German Army, though it had lost its trench system, was not yet broken. Its rearguards were armed with thousands of machine-guns, and they made impossible the rapid pursuit by cavalry which would have turned retreat to rout. Only the tanks could have done that; two brigades of them might have done it; but now after three months' continuous fighting very few remained, and the corps itself had lost a third of its personnel.

The steady attack, the methodical pursuit continued, and a week later, on October 17th, one brigade of tanks went into action with the 4th Army in a combined British and French attack, south of Le Cateau. The numerous waterways of this flat country were now the chief defence of the Germans. At this part of the line they were protected by the river Selle, which lay between the two armies. To guide the attacking tanks, tape was laid across the river at night time, and it was then discovered that the Germans had dammed the stream in places to increase the difficulties.

The morning of the 17th came with such a heavy fog that the

Tanks moved up to the attack by compass bearings. Each tank of the twenty-eight carried a "Crib," [1] and with the help of these the river was crossed.

The enemy made little resistance. They had trusted to the streams. Three days later tanks again crossed the Selle in an attack north of Le Cateau. This time the engineers built a bridge for them. It was built at night and was just beneath the surface of the water, so that it was invisible to the enemy by day. By this underwater bridge all the tanks safely crossed.

Two days later thirty-seven tanks took part in a moonlight attack. It was successfully carried through in spite of mist and heavy gas shelling, and the infantry found the tanks as useful in making a way for them through the hedges of the unfortified country as through the wire belts of the trench zones.

On November 4th, French and British together attacked on a thirty-mile front from Valenciennes to the river Oise, and thirty-seven tanks worked with the British infantry. That day two supply tanks also joined in the battle near Landrecies. They were carrying forward bridging materials, when they found the infantry, still on the western bank of the canal and unable to advance because of the German machine-guns. Supply tanks are not meant for fighting, but these two went at once into action, and the infantry followed them as if they were fighting tanks. The machine-gunners surrendered. The canal was crossed.

The next day eight Whippets supported the 3rd Guards Brigade in a successful attack across a difficult country of fences and ditches north of the Forest of Mormal. This was the last action that the tanks fought. The corps was now so reduced that companies had taken the place of battalions and sections the place of companies.

Six days later, when hostilities ceased, the corps was busy trying to reorganise a fighting force out of its diminished and weary battalions.

Since the 430 tanks began the battle of Amiens on August 8th, there had been ninety-six days of almost continuous fighting. On thirty-nine days out of the ninety-six the Tank Corps had been engaged, and 1993 tanks and tank armoured cars had gone into action. Eight hundred and eighty-seven had been handed over to salvage, but only

---

1. This consisted of a strong hexagonal-shaped framework of timber braced with steel members. It was 5 feet across and 10 feet long, and was used in the same way as the "Fascine," but its weight was only 12 cwt., as compared with the 30 cwt of the "Fascine."

fifteen of these had been struck off the strength as altogether beyond repair, while 214 had been repaired and returned to their battalions.

The casualties, compared with the strength of the corps, had been heavy. Five hundred and ninety-two officers were killed, wounded, prisoners or missing out of a total of 1500, and 2562 other ranks out of a total of 8000. Yet by the standard of the casualties of the infantry in those battles which they fought unsupported by tanks, these losses were small indeed. The whole corps was less than the strength of an infantry division, and in those thirty-nine days of fighting, in which victorious armies had been completely broken, it had lost less than many infantry divisions, during the battle of the Somme, had lost in a single day.

It is not possible to compare those figures with the losses that the infantry would have sustained had they had to do the work of the tanks, for the tanks continually did what it was beyond the power of flesh and blood to do. One can only say that if the infantry had been able to do it at all they would have paid a price in lives many hundred times as great.

What did the Germans think of the tanks? It is credibly reported that when Hindenburg visited the German tank centre near Charleroi in February 1918 he said, "I do not think that tanks are any use, but as these have been made they may as well be tried." That he said this was certainly believed in the German Tank Corps, which was not much encouraged thereby, and if he said it he only repeated what Lord Kitchener had said of our tanks three years before when he first saw them at Hatfield.

Other German generals believed in them if Hindenburg did not, and the commander of the 17th German Army said of them: "Our own tanks strengthen the *morale* of the infantry to a tremendous extent even if used only in small numbers, and experience has shown that they have a considerable moral effect on hostile infantry."

The great Allied attack had only just begun when the German Government showed that it recognised the growing danger of the new weapon. Speaking in the Reichstag for the minister of war, at the time of the battle of Amiens, General von Wrisberg said, "The American Armies need not terrify us. We shall settle with them. More momentous for us is the question of tanks."

Then just before the end this message from the Prussian Minister of War was sent out:

The superiority of the enemy at present is principally due to their use of tanks. We have been actively engaged for a long time in working at producing this weapon (which is recognised as important) in adequate numbers. We shall then have an additional means for the continuance of the war if we are compelled to continue it.

So one of the last efforts to hearten the German people was a promise of tanks.

But it is not with any reluctant tribute from a German that I wish to end this story of how we built the tanks. I have already quoted the British commander-in-chief's first words on them:

> Wherever the tanks advanced we took our objectives, and where they did not advance we failed to take our objectives.

His last words, in his dispatch of December 21st, 1918, are these:—

> Since the opening of our offensive on 8th August, tanks have been employed in every battle, and the importance of the part played by them in breaking the resistance of the German infantry can scarcely be exaggerated. The whole scheme of the attack of the 8th of August was dependent upon tanks, and ever since that date on numberless occasions the success of our infantry has been powerfully assisted or confirmed by their timely arrival. So great has been the effect produced upon the German infantry by the appearance of British tanks that in more than one instance, when for various reasons real tanks were not available in sufficient numbers, valuable results have been obtained by the use of dummy tanks painted on frames of wood and canvas.

> It is no disparagement of the courage of our infantry or of the skill and devotion of our artillery to say that the achievements of those essential arms would have fallen short of the full measure of success achieved by our armies had it not been for the very gallant and devoted work of the Tank Corps, under the command of Major-General H. J. Elles.

What we had claimed that the tanks could do they had done.

CHAPTER 11

In Conclusion

"LET US NOW PRAISE FAMOUS MEN"

One of the results of the secrecy which had to be maintained about mechanical warfare is that the public knows nothing of those men whose genius designed tanks, whose enthusiasm and energy compelled a doubting and reluctant War Office to use them, and whose skill in the field made them a terrifying weapon.

The following are the men whom Great Britain has to thank for its tanks, and for the honour of having given to the Allied cause the greatest invention of the war:—

Mr. Winston Churchill, who first encouraged in a definite way the new idea of mechanical warfare by appointing a committee of the Admiralty to study it and by authorising the funds for the Admiralty to develop it.

Mr. Lloyd George, who protected the idea from destruction by the forces of doubt and reaction on many occasions.

Mr. Edwin S. Montagu, who succeeded Mr. Lloyd George as minister of munitions and worked for mechanical warfare in the same spirit of enthusiasm.

The Admiralty, which encouraged the whole development. Without its support and help tanks would never have been produced.

Sir Eustace d'Eyncourt, who was the real father of the tanks and nursed the development from the beginning to the end. He on all occasions gave his great technical knowledge and experience and the weight of his personal influence without fear and without stint.

Major W. G. Wilson and Sir William Tritton, who brought all their experience, energy, mechanical knowledge and inventive genius to the

ERNEST SQUIRES ESQ.

working out of the mechanical details of the tanks.

Major N. E. Holden, who was my loyal and reponsible deputy and really did all my work for three years.

Sir Sigmund Dannreuther, Director of Finance, who in his important position gave mechanical warfare invaluable help. His great intelligence and broad-minded views helped the development from the beginning. Numerous difficult problems, both in the mechanical warfare department and Anglo-American commission, were submitted to him and always solved.

Major K. P. Symes, who showed untiring energy and skill in the development and production of light armour-plate without which the tanks would have proved of little value, and superintended, with Lieut. W. E. Rendle, the whole production of tanks.

Major-General E. D. Swinton, who gave the War Office not one moment's peace until they had adopted this new method of warfare, and who raised the first tank force and commanded it in the first tank battle in September 1916.

The Royal Tank Corps, whose magnificent courage and *esprit de corps* were second to none in any army in the world.

Major-General H. J. Elles, who succeeded General Swinton and led the Corps into action at Cambrai, and, with his chief of staff, Colonel J. F. C. Fuller, was responsible for the tactics, efficiency and magnificent *esprit de corps* of the Tank Corps.

General Scott Moncrieff, who was appointed chairman of the landship committee when Mr. Churchill left the Admiralty.

Lieut. Percy Anderson, who from the earliest days was responsible under me for the whole of the organisation of the mechanical warfare department.

Mr. L. W. Blanchard, who was head of my drawing-office, and showed not only his exceptional technical ability and experience, but untiring energy under most difficult circumstances.

Lieut. R. A. Robertson, chief inspector of all tank production.

Lieut. R. Spinney, his deputy.

Captain T. L. Squires, who was responsible for the construction of gun-carrying machines, and who trained the crews who took the first of this type of tank into action.

Squadron 20, R.N.A.S., and Commander McGrath, who carried

out all testing, experimenting and transport of the tanks until the end.

Sir Charles Parsons, who gave invaluable help as technical adviser to the mechanical warfare department.

Mr. Dudley Docker, Major Greg, Mr. Lincoln Chandler, Mr. E. Squires and Mr. Stockton of the Metropolitan Carriage Wagon and Finance Company, who organised the production of tanks on that large scale without which they could never have played their decisive part.

Sir William Beardmore and Mr. T. M. Service, who supplied armour and bomb-proof plates.

Sir Robert Hadfield, to whom we owed the special steel which gave the tanks their length of life and trustworthiness.

Mr. Morgan Yeatman, a genius in bridge building, who was technical adviser on stresses and strains in this novel landship.

Mr. Starkey of Messrs. William Foster & Company of Lincoln, head draughtsman under Sir William Tritton, who was responsible for the drawings of the first experimental and first successful tank.

Mr. Sykes, works manager of Fosters, who produced the experimental tanks and was building them from that time onwards until the Armistice.

Sir George Hadcock and Mr. H. L Brackenbury of Messrs. Armstrong, Whitworth and Company, who designed the guns and gun-mountings, and never failed us in producing them in enormous numbers when the general opinion was that such production was impossible.

Messrs. The Daimler Company and Mr. Percy Martin, who never failed in the supply of engines up to time.

Engine Patents, Ltd., and Mr. H. Ricardo, who designed and produced the Ricardo engine in thousands.

All the manufacturers, works managers, foremen and workmen who, at all times, willingly worked night and day. I never remember a single instance of a strike in any of our tank factories. No country could be prouder than this country should be of the work done by the men who built the tanks.

Nor would the story of the tanks be complete without a tribute to the work done by the women. There was hardly any part of the tank upon which women were not employed before the end of the war.

In my own department. Miss L. P. Pérot was appointed assistant director. This brought forth from the establishment branch of the Ministry of Munitions a protest that such an appointment had never been made before and therefore could not be approved. But the appointment was made and it was justified. To her and the thousands of women who worked for the tanks the country owes its gratitude, as well as to the men.

Appendices

Appendix 1

LETTER FROM MR. ASHMEAD BARTLETT

> Marlborough Chambers,
> Jermyn Street,
> St. James's S.W.,
> March 5th, 1918.

Dear Stern,

I am sending you the promised relic—namely, the (Pyrene) fire extinguisher of the tank, H.M.S. *Corunna*, which I recovered from her interior on October 7th, 1916. I think I was the first person to ever write a description of the tanks which was allowed to appear in the press. I will now quote from my diary of that date:—

I was endeavouring to get into the village of Combles, but was stopped by heavy shell-fire. I was taking cover in a trench on the ridge overlooking Combles, when I suddenly espied a strange-looking object lying close to the Bois de —— (I cannot remember the name of this wood, and have no map with me for the moment), about 150 yards away. I soon discovered with the aid of my glasses that it could only be a mortally wounded tank. I then crawled over the shell-shattered ground to the spot. All around me were evidences of a fearful *mêlée*; the dead, still unburied, lay in hundreds.

Some of the corpses had been partly covered with the earth which the heavy rains had washed away, exposing the grinning, half-decayed, blackened faces of the dead to view. Around the Tank itself they lay thicker than ever, showing how our infantry had attacked in her wake, endeavouring to obtain some shelter from the murderous machine-gun fire by crawling behind her. The Tank itself, H.M.S. *Corunna*, was lying totally disabled, having been knocked out by a shell which had smashed the engine; but as there were no dead inside, I do not know what became of the crew. It seemed that this landship had reached the edge of the wood and made a counter-attack, for our dead and the German dead lay pell-mell round this iron monster. I crawled inside and found the interior a mass of mangled machinery, cartridge belts and Hotchkiss shells. . . . . The curious part of visiting H.M.S. *Corunna* was this—I was able to write a description of the tanks which the French military censor immediately passed, so that the first account of them did not come from the correspondents attached to the British Army, but from those attached to the French. Passing along the ridge from Combles to Hardicourt, I discovered another of our dying tanks which had fallen in a gallant attack on the village. The shell-fire was too heavy for me to approach, so I could collect no souvenirs from it.

        Yours very sincerely
            Ashmead Bartlett.

Appendix 2

PROGRAMME OF TANK DISPLAY AT OLDBURY ON MARCH 3RD, 1917

SECRET.

MINISTRY OF MUNITIONS.	MINISTERE DES MUNITIONS.
MARCH 3RD, 1917.	LE 3 MARS, 1917.
Demonstration	Séance de Démonstration
OF	DE
EXPERIMENTAL TANKS.	**DIVERS TYPES DE TANKS.**
Owing to the Confidential Nature of this Pamphlet, visitors are particularly requested to return it to the Officer stationed at the Gate before leaving.	Ce Programme est confidentiel, et il est particulierement recommandé de le remettre à la sortie à l'Officier de Garde.
A. G. STERN, Lieut.-Colonel, *Director-General W.M.S.D.*	A. G. STERN, Lieut.-Colonel, *Directeur-General M.W.S.D.*
Particulars of Arrangements.	**PROGRAMME.**
The Tanks will line up in front of the Stand for General Inspection and Examination.	Les "Tanks" seront assemblées en face du Pavillon pour une inspection générale.
The Tanks will leave for end of field, and having lined up there, will start simultaneously and cross the trenches, finally returning to the starting-point.	Les "Tanks" se mettront en marche pour un rassemblement à l'extrémité du champ, au point de départ. Partant de ce point les "Tanks" feront la traversée des tranchées et reviendront au point de départ.
Various Tanks will navigate the large Shell Holes in front of the Stand.	Plusieurs "Tanks" feront la traversée des "entonnoirs" en face du Pavillon.

Visitors are requested to leave the Ground not later than 3 o'clock.
The special train for London will leave at 3.15 p.m.

Les Visiteurs sont priés de quitter le champ au plus tard a 3 hr.
Le train pour Londres partira a 3 hr. 15.

Types	Distinguishing colours
1. Original Standard Machine	BLACK (Noir).
2. Tritton Chaser	GREEN (Vert).
3. Williams-Janney Hydraulic	SKY BLUE (Azur).
4. Wilson Epicyclic	WHITE (Blanc).
5. Daimler Petrol-Electric	RED (Rouge).
6. Westinghouse Petrol-Electric	YELLOW (Jaune).
7. Wilkins's Multiple Clutch	PINK (Rose).
8. Gun-carrying Machine	

LIST OF OFFICIALS PRESENT AT DEMONSTRATION AT OLDBURY, MARCH 3, 1917.

GENERAL ANLEY.
LIEUT. ANDERSON.
M. BRETON.
GENERAL BINGHAM.
GENERAL BUTLER.
COMMANDER BORIS.
LIEUT.-COMMANDER BARRY.
MAJOR BROOKBANK.
M. BRILLIE.
LIEUT. BROGLIE.
M. CROCHAT.
M. DE LA CHAUME.
CAPTAIN CHARTERIS.
LIEUT. CUDDY.
GENERAL DESSINO.
LIEUT.-COLONEL DESLANDRES.
LIEUT.-COLONEL DOUMENO.
CAPTAIN DEPOIX.
M. DUPONT.
A. DAY, ESQ.
DUDLEY DOCKER, ESQ.
DUDLEY DOCKER, JNR.
GENERAL ESTIENNE.
SIR E. D'EYNCOURT.
COMMODORE EVERETT.
SIR L. WORTHINGTON EVANS.
COLONEL ELLES.
CAPTAIN EDWARDS, R.N.

LIEUT. BRANDON.
L. W. BLANCHARD.
CAPTAIN BUSSELL.
LIEUT. W. BRAY.
CAPTAIN BOULTON.
PAY-MASTER BIRD.
COLONEL COURAGE.
C. M. CARTER, ESQ.
DUGALD CLERK, ESQ.
LIEUT.-COLONEL CHALLEAU.
COLONEL MILMAN.
COMMANDER MCGRATH.
J. MASTERTON SMITH, ESQ.
C. H. MERZ, ESQ.
M. MANTOUX.
CAPTAIN MICHEL.
LIEUT.-COMMANDER NOGUES.
COLONEL NOOT.
GENERAL VICOMTE DE LA PANOUSE.
COLONEL LORD PERCY.
SIR C. PARSONS.
LIEUT.-COMMANDER PERRIN.
LIEUT. ROBERTSON.
LIEUT. RENDLE.
H. R. RICARDO, ESQ.
GENERAL SIR HERBERT SMITH.
COMMANDANT SENESCHAL.

Major Garrick.
Major Greg.
A. P. Griffiths, Esq.
Lieut. Grossmith.
Lieut. Gordon.
Captain Hope, R.N.
Colonel Sir M. Hankey.
Colonel Haynes.
Captain Holden.
Captain Hatfield.
G. H. Humphries, Esq.
Lieut. Hubert.
General de Jongh.
Captain de Jarney.
Colonel James.
General Kiggell.
Comte de Chasseloup Laubat.
Colonel Lannowe.
Captain Leisse.
Colonel Hardross Lloyd.
Lieut. R. Levaique.
General F. B. Maurice.
Sir E. Moir.
M. Sabatier.
Captain Sanderson.
Captain Symes.
Colonel Symon.
Colonel Stern.
Lieut. Shaw.
Captain Stevens.
Colonel Searle.
F. Skeens, Esq.
E. Squires, Esq.
Sir W. Tritton.
Captain Trelawnay.
Lieut. Thornycroft.
P. Turner, Esq.
F. J. Todd, Esq.
Captain Vyvyan, R.N.
Lieut.-Commander Velpry.
General Sir R. D. Whigham.
Sir Glyn West.
Sir John Weir.
A. Wilson, Esq.
Major Wilson.
Lieut. Weston.

TABULATED DATA

Machine.	H.P.	Speeds in M.P.H.				Total Weight.	Track Pressures in Lbs. per Sq. In.		H.P. Weight Ratio.
		1.	2.	3.	4.	Tons.	Max.	Min.	Lbs. per H.P.
No. 1	105	¾	1½	2	4	26	23·1	5·8	555
No. 2	100	1¾	2½	4	7½	12	13·6	3·2	268·8
No. 3	105	Variable to 4				28	24·9	6·2	597
No. 4	105	¾	1½	2	4	26	23·1	5·8	555
No. 5	125	Variable to 4				28	24·9	6·2	501·8
No. 6	115	Variable to 4				28	24·9	6·2	545·4
No. 7	105	1	1·9	3·9	—	26	23·1	5·8	555
No. 8	105	¾	1½	2	4	34	35·5	6·0	725

STANDARD MARK IV.

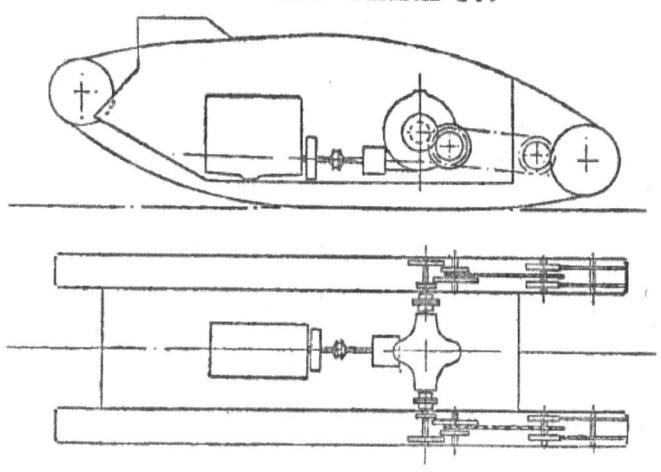

Standard Machine No. 1.

Description. — This is the Standard machine that has been used in France, with the exception of a few minor modifications and the discarding of the tail.

Power Unit. — One 6 cylinder 105 B.h.p. Daimler engine, running at 1000 r.p.m.

Transmission. — The drive is taken from the engine by a clutch, through a two-speed and reverse gear box, to worm gear and differential driving a cross shaft; thence it is taken to each track by independent two-speed gears and chain to a counter shaft on which are mounted two gear wheels engaging with the track driving sprockets

Type Normal.

Ce type de machine est celui dont nous nous sommes servis depuis son introduction, mais avec quelques modifications, dont la supression des deux roues directrices est la plus importante.

Le moteur est à six cylindres de cent cinq chevaux à mille tours par minute. Système Daimler.

Transmission. — La puissance du moteur passe par moyen d'un embrayage ordinaire et un changement de vitesse double avec marche en arrière, à un engrenage à vis sans fin et avec différentiel, dont l'axe est disposé à travers la machine. Cet axe porte à chaque bout deux roues dentées formant un changement de vitesse à double effet. De là,

Speed Control.—This transmission provides for four speeds forward and two reverse, but necessitates the stopping of the machine in order to change gear.

Steering.—The steering is carried out by braking one or other side of the cross shaft when the differential lock is out, or by the independent use of the secondary two-speed gears.

Estimated Speeds.—
1st gear, ¾ of a mile per hour.
2nd   „   1½ miles per hour.
3rd   „   2   „   „
4th   „   4   „   „

Armour.—The machine is protected by armour-plate varying from 6 to 12 mm. in thickness.

Armament.—This consists of either of two short 6-pounder 23 calibre Q.F. guns, and four Lewis guns, or of six Lewis guns.

Weight.—The weight of this machine complete is about 26 tons.
Weight, horse power ratio, 555 lbs. per h.p.

Track Pressure.—
Maximum (5 ft. ground line) 23·1 lbs. per sq. in.
Minimum (20 ft. ground line) 5·8 lbs. per sq. in.

une chaîne transmet l'effort à deux roues dentées qui s'engagent avec les roues dentées des deux chenilles qui se trouvent une à chaque côté de la machine. Cela donne a quatre vitesses en avant, et deux en arrière, mais l'arrêt du mouvement de la machine est necessaire pour le changement de vitesse.

Direction.—La direction se fait par le freinage de la chenille d'un côté ou de l'autre pendant que le différentiel est debloqué, ou par l'emploi indépendant des deux changements supplémentaires, avec le différentiel bloqué.

Vitesses Calculées.—
1re Vitesse 1·206 kilomètres a l'heure.
2me Vitesse 2·418 kilomètres a l'heure.
3me Vitesse 3·219 kilomètres a l'heure.
4me Vitesse 6·437 kilomètres a l'heure.

Blindage.—Le blindage est de 6 mm. à 12 mm. selon la position.

Armement.—Il y a deux types de cette machine dont un est armé de deux petits canons à tir rapide portant un obus de 2·724 kilos, et deux mitrailleuses Lewis, et l'autre, de six mitrailleuses Lewis seulement.

Le poids complet est de 26,390 kilos.
Poids du Tank pour un cheval vapeur, 252 kilogs.
Lorsque les deux chenilles portent sur une longueur d'environ 1 m. 50, le poids pour centimètre carré est de 1 kilog. 600. Enfoncée dans la boue le poids pour centimètre carré est de 400 grammes environ.

Maximum poids supporté par la chenille d'une longueur 1·524 metres sur terre dure par centimètre carré 1·6 kilos. Enfoncée dans la boue jusqu'à 6·1 mètres ·4 kilos par centimètre carré.

TRITTON CHASER.

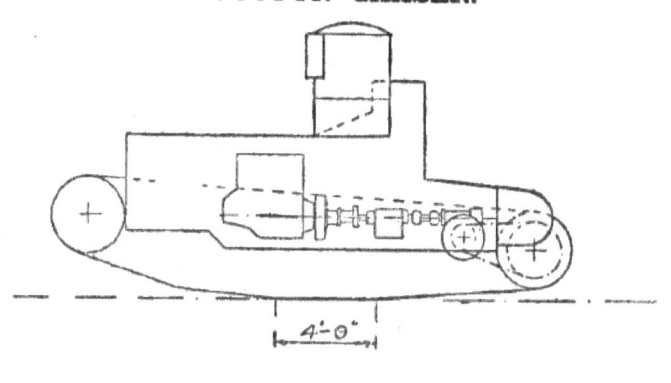

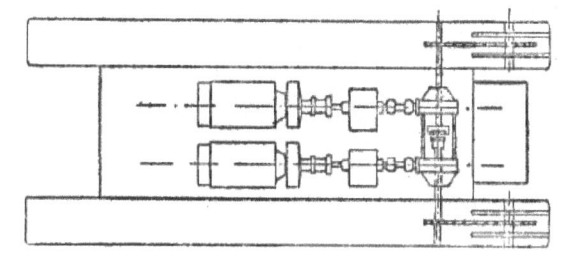

|— 4'-0" —|

Tritton's Light Machine, No. 2 (E.M.B.)

Description.—This machine is an attempt at meeting the requirements of the military authorities for a light machine capable of maintaining higher speeds than the Standard machine. Its twin engine transmission is worthy of note.

Power Unit.—The power unit consists of two 4-cylinder 50 h.p. Tylor engines.

Transmission.—(One set for each track.) The drive is taken from the engine by a clutch and four-speed and reverse sliding gear box to worm gear driving a divided cross shaft, thence by chain to the rear driving sprocket as in the Standard machine. A differential lock is provided by sliding dog clutches on the divided cross shaft.

Speed Control.—The gear on this machine can be changed while

Machine Legère. Système Tritton.

Cette machine est un essai pour remplir toutes les conditions demandées par l'état-major pour une vitesse de marche plus considérable que celle du type normal.

Transmission.—La transmission à moteurs jumeaux est digne d'attention. Ces moteurs sont à quatre cylindres, système Tylor, d'une puissance de 50 chevaux chaque. Chaque transmission se fait par embrayage et changement normal à quatre vitesses et avec vis sans fin. Les deux arbres de ces engrenages sont disposés transversalement la machine, et sont engagés comme dans la machine normale, avec les chenilles. Il y a un manchon pour accoupler les arbres, ce qui reproduit le différentiel de la machine normale. Il est possible de changer la vitesse de la ma-

the machine is in motion, by operating first one gear and then the other.

Steering.—The steering is carried out by throttle control of the two engines. The engine on the side opposite to that to which it is desired to turn is accelerated, and the other throttled.

Estimated Speeds.—
1st gear, 1¾ miles per hour.
2nd „   2¼   „   „
3rd „   4   „   „
4th „   7½   „   „

Armour.—The vital portions of this machine are armoured with 9 mm. plate, those not directly exposed to enemy's fire being protected by 6 mm. plate.

Armament.—The armament consists of one Lewis gun in a gun turret.

Weight.—The weight of this machine complete is about 12 tons.

Weight, horse power ratio, 268·8 lbs. per h.p.

Track Pressure.—
Maximum (4 ft. ground line) 13·6 lbs. per sq. in.
Minimum (17 ft. 2 in. ground line) 3·2 lbs. per sq. in.

chine sans arrêt au moyen des deux changements, qu'on doit manipuler l'un après l'autre.

Direction.—La direction se fait par l'intermédiaire des deux soupapes à papillons qui forment un mécanism à actions solidaires. L'accélération du moteur à gauche fait tourner la machine à droite, et vice versâ.

Vitesse de marche.—
1re Vitesse 1¾ milles par heure = 2·815 kilomètres.
2me Vitesse 2¼ milles par heure = 4·023 kilomètres.
3me Vitesse 4 milles par heure = 6·437 kilomètres.
4me Vitesse 7½ milles par heure = 12·079 kilomètres.

Blindage.—Les parties vitales sont protégées par un blindage de 9 mm. et les autres parties par un de 6 mm.

Armement.—Une mitrailleuse Lewis est montée dans une tourelle au-dessus de la machine.

Poids.—9,136 kilos.
Poids par cheval, 90·8 kilos.
Pour une portée de 1 m. 15 le poids est de 76 kilogs. par centimètre carré de chenille qui est reduit à 175 grammes lorsque la portée 5 met. 25.

WILLIAMS-JANNEY HYDRAULIC GEAR.

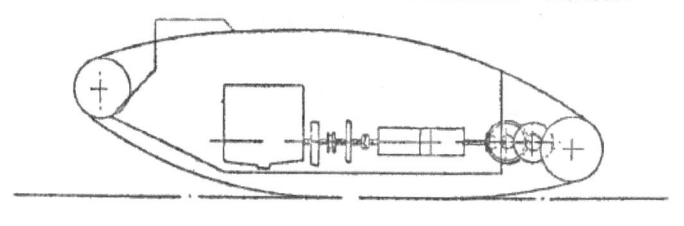

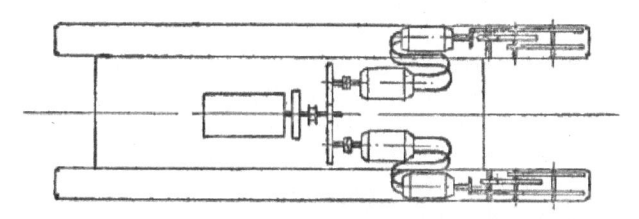

Williams-Janney Hydraulic Machine, No. 3.

Description.—This machine is fitted with the Williams-Janney hydraulic transmission, in which the working fluid is oil. The pumps and motors work on the same principle, being of rotary type with cylinders parallel to the axis of rotation. Reciprocating motion is given to the pistons by means of a disc whose perpendicular axis may be inclined to the shaft. By varying the inclination of this disc to the shaft, the stroke and consequently the delivery of the pumps may be altered as occasion demands. The stroke of the motors is invariable.

Power Unit.—The power unit consists of single six-cylinder 105 B.h.p. Daimler engine.

Transmission.—This is hydraulic. The two pumps are arranged side by side and are driven from the engine by means of a reduction gear, giving a four-to-one reduction. The two motors transmit the drive through bevel gear and a single spur reduction gear to a wide gear wheel arranged in between the track-driving sprockets.

Système Williams-Janney.

La machine fonctionne par l'intermédiaire de l'huile sous pression. Les pompes et les moteurs, dont il y a une paire pour chaque chenille sont identiques. Les cylindres sont disposés autour, et avec leurs axes parallels à l'axe de revolution de l'essieu. Le mouvement va et vient des pistons de la pompe se produit par l'inclinaison, d'un diaphragme sur lequel aboutent les bielles.

Le changement de l'angle de ce diaphragme controle la quantité d'huile livrée par la pompe, car la course des pistons est variable, et controle aussi le sens de la marche, et les tours par minute de l'axe du moteur à huile sous pression.

Moteur.—Le moteur à essence est de 105 chevaux Daimler.

Transmission.—La transmission se fait comme il est deja dit. Les pompes tournent à ¼ de vitesse, c'est-à-dire à 250 tours. Les Moteurs hydrauliques sont engagés par l'intermediaire des roues coniques avec les roues dentées qui s'engagent à leur tour avec les chenilles.

Contrôle.—La vitesse est controlé par l'inclinaison du dia-

Speed Control.—The speed control is infinitely variable and is carried out by varying the inclination of the discs in the pumps.

Steering.—The steering is carried out by increasing the stroke of the pump on the side opposite to that to which it is desired to turn.

Estimated Speeds.—The speed is infinitely variable from zero to four miles per hour.

Armour.—Same as No. 1.

Armament.—Same as No. 1.

Weight.—The weight of this machine complete is about 28 tons.

Weight, horse power ratio, 597 lbs. per h.p.

Track Pressure.—
Maximum (5 ft. ground line) 24·9 lbs. per sq. in.
Minimum (20 ft. ground line) 6·2 lbs. per sq. in.

phragme. La direction se fait par l'accélération d'une chenille ou de l'autre, moyennant l'inclinaison des diaphragms des pompes.

Vitesse.—La vitesse varie de zero jusqu'a 6·437 kilomètres à l'heure.

Blindage, et Armement comme la machine normale.

Poids.—Environ 2,420 kilos.

Proportion 271 kilos par cheval.

Pression par centimètre carré.
Maximum 1·75 kilos.
Minimum ·45 kilos.

WILSON'S EPICYCLIC TRANSMISSION.

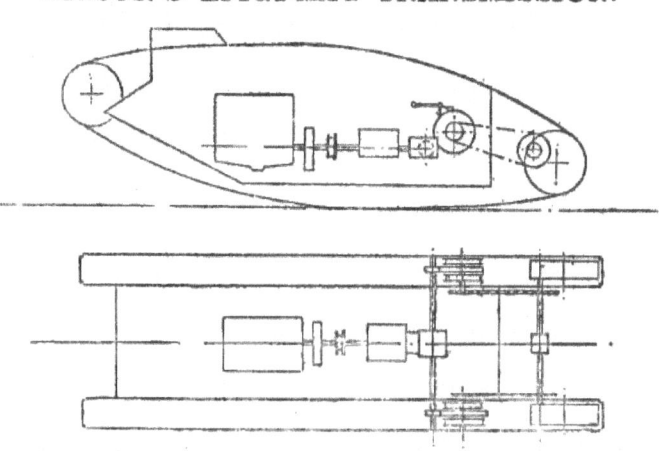

Wilson's Epicyclic Gear Machine, No. 4 (E.M.E.)

Description.—This transmission has been designed to give as its chief object better steering control, and secondly, to do away with the large worm reduction gear as used on the Standard machine. The transmission is epicyclic only in so far as the epicyclic principle is employed to give good steering control without the use of a clutch, combined with reduction gearing.

Power Unit.—The power unit consists of a single six-cylinder 105 B.h.p. Daimler engine.

Transmission.— The drive is taken from the engine by a clutch and four-speed and reverse sliding gear box to a bevel driven cross shaft; thence it is taken to each track,

through an epicyclic reduction gear independently operated by means of a brake, and by chain, to a divided cross shaft on which is mounted a pinion engaging with a central gear wheel between the road chain driving sprockets. A differential effect is obtained by means of dog clutches on the divided cross shaft.

Speed Control.—This transmis-

Système Planétaire Wilson.

Ce système est destiné premièrement à augmenter la facilité du contrôle, et à supprimer la grande caisse avec roue hélicoïdale de la machine normale.

Le système planétaire est employé seulement pour la facilité de contrôle, et pour la combinaison avec une réduction de vitesse rendue possible et non pour un changement de vitesse.

Le moteur est de six cylindres 105 chevaux Daimler.

Transmission.—La transmission se fait par le moyen d'un embrayage ordinaire et un changement à quatre vitesses; un engrenage à roues coniques, où se trouve l'appareil pour la marche en arrière (ce qui donne

par conséquence quatre vitesses en arrière), et enfin, pàr un essieu transversal. A chaque bout de cet essieu se trouve une roue dentée, qui s'engage avec l'anneau extérieur du système planétaire, une chaîne communique l'effort à la chenille avec quelques modifications de la machine normale.

Le différentiel est supprimé et

Speed Control.—This transmission provides four forward speeds and one reverse, to operate any of which it is necessary to stop the machine.

Steering.—The steering is carried out by operating independently the epicyclic reduction gear brakes.

Estimated Speeds.—
1st gear, ¾ of a mile per hour.
2nd ,, 1½ miles per hour.
3rd ,, 2 ,, ,,
4th ,, 4 ,, ,,

Armour.—Same as No. 1.

Armament.—Same as No. 1.

Weight.—The weight of this machine complete is about 26 tons.

Weight, horse power ratio, 555 lbs. per h.p.

Track Pressure.—
Maximum (5 ft. ground line) 23·1 lbs. per sq. in.
Minimum (20 ft. ground line) 5·8 lbs. per sq. in.

Le differentiel est supprimé et toute la puissance du moteur se porte sur l'autre chenille. C'est dans cette manière que la direction se fait.

Vitesse Calculée.—
1re Vitesse 1·206 kilomètres par heure.
2me Vitesse 2·1 kilomètres par heure.
3me Vitesse 4·0 kilomètres par heure.
4me Vitesse 6·8 kilomètres par heure.

Blindage.—Armement et Poids les mêmes que dans la normale.

DAIMLER PETROL-ELECTRIC.

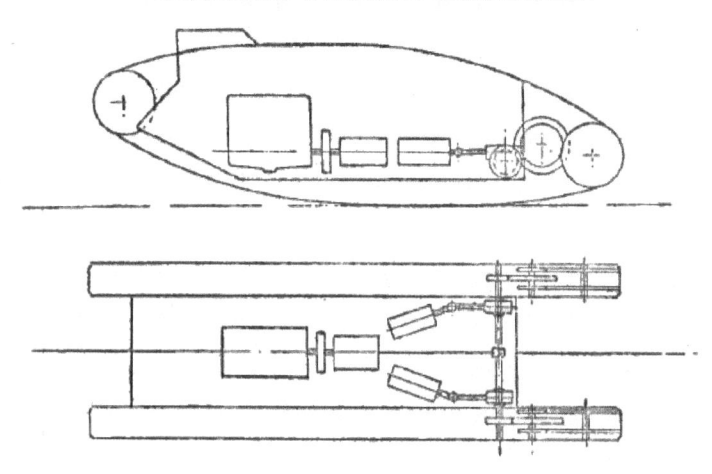

Daimler Petrol-Electric Machine, No. 5 (E.M.F.)

Description.—The chief feature of this transmission is the entire absence of controllers and external wiring, except for the main leads from the generator to the motors. The control is entirely effected by shifting the brush position on the generator and motors, arrangements being made to prevent excessive sparking.

Power Unit.—This consists of a six-cylinder Daimler engine fitted with aluminium pistons and a lighter flywheel. The normal speed is 1400 r.p.m.

Transmission.—A single generator coupled direct with the engine supplies current to two motors in series. The independent control of each motor is accomplished by shifting the brushes. Each motor drives through a two-speed gear box to a worm reduction gear and from thence through a further gear reduction to the sprocket wheels driving the road chain driving wheels. A differential lock is obtained by connecting the two worm wheel shafts by a dog clutch.

Speed Control.—The speed is in-

Machine electrique Daimler.

Le trait le plus remarquable de cette transmission est la suppression du controle et des fils extérieurs, sauf les conducteurs entre dynamos et moteurs.

Controle.—Le contrôle est entièrement gouverné par le mouvement des balais autour de leurs collecteurs, et sans aucun crachement au balais.

Le moteur est à essence avec pistons en aluminium, et un volant plus léger. La vitesse normale est de 1,400 tours par minute.

Transmission.—Une dynamo accouplée au moteur fournit l'énergie à deux moteurs electriques en série. Le contrôle independant des moteurs électriques se fait par le mouvement des balais.

Chaque moteur fait marcher, par l'intermediaire d'un double changement de vitesse, une réduction à vis sans fin, et encore une réduction avec les roues dentées de la chenille.

L'action différentielle est obtenue par la deplacement d'un manchon d'accouplement entre des deux vis sans fin.

finitely variable within limits and is controlled by shifting the brushes on the generator and motors, the engine running at a governed speed of 1400 r.p.m.

Steering.—The steering is carried out by means of the brush control on the motors.

Estimated Speeds.—This varies from zero to four miles per hour.

Armour.—Same as No. 1.

Armament.—Same as No. 1.

Weight.—The weight of this machine complete is about 28 tons.

Weight, horse power ratio, 501·8 lbs. per h.p.

Track Pressure.—
  Maximum (5 ft. ground line) 24·9 lbs. per sq. in.
  Minimum (20 ft. ground line) 6·1 lbs. per sq. in.

La vitesse varie de zero à 6·437 kilomètres.

La direction se donne par le changement de vitesse des deux chenilles, selon l'explication donnée ci-dessus.

Blindage.—Armement, comme le type normale.

Poids.—28,420 kilos.

Proportion, 227 kilos par cheval.

Pression par centimètre carré.
  Maximum 1·70.
  Minimum ·43.

BRITISH WESTINGHOUSE PETROL-ELECTRIC.

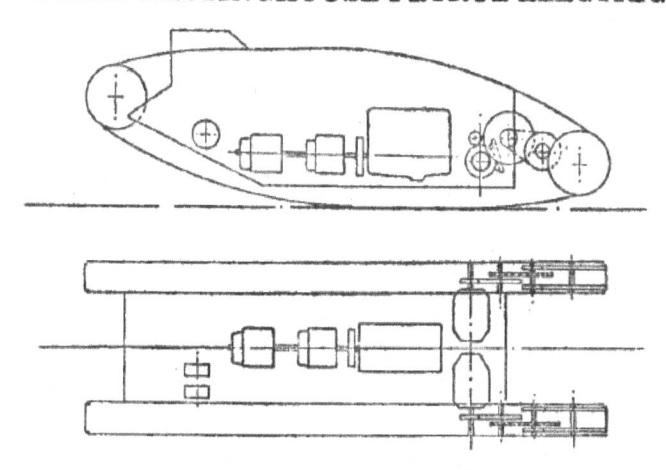

British Westinghouse Petrol-Electric Machine, No. 6 (E.M.G.)

Description.—In this transmission two separate sets consisting of generator and motor are used, the two generators being driven in tandem from the forward end of the engine; one exciter being used for both generators.

Power Unit.—The power unit consists of the Standard six-cylinder Daimler engine, running at 1200 r.p.m.

Transmission.—Each generator provides current for its own motor, which drives its track by double reduction spur gear and chain drive to a counter shaft, on which is mounted a pinion engaging with a gear wheel mounted between the two track sprockets.

Speed Control.—The speed control is infinitely variable between limits, and is effected by means of rheostats controlling the exciter currents to the fields, reversing being carried out by separate reversing switches interlocked in such a way that the current must be cut off before the switches can be operated.

Steering.—The steering is carried

Machine electrique British Westinghouse.

Dans cette machine il y a deux installations indépendantes composées chacune d'un dynamo et un moteur, pour chaque chenille. Une seule dynamo excitatrice sert pour les deux installations. Les deux dynamos sont installées en "serie," devant le moteur Daimler à 105 chevaux, qui est semblable à celui de la machine normale. Chaque dynamo transmet l'energie à son propre moteur et par l'intermédiaire d'une double reduction à roues dentées, et d'une chaîne, l'effort est transmis à la roue dentée de la chenille.

Le contrôle variable est entre zéro et 6·437 kilomètres et se fait par des rheostats contrôlant les champs électriques des moteurs. La marche en arrière est controllée par un manipulateur inverseur indépendant, à actions solidarisées, de manière qu'il faut que le courant soit interrompu avant que l'opération d'inversion se fasse. La direction se fait par l'opération indépendante des moteurs.

Blindage.—Armement comme

out on this machine by operating the motors independently.

Estimated Speeds.—This varies from zero to four miles per hour.

Armour.—Same as No. 1.

Armament.—Same as No. 1.

Weight.—The weight of this machine complete is about 28 tons.

Weight, horse power ratio, 545·4 lbs. per h.p.

Track Pressure.—

Maximum (5 ft. ground line) 24·9 lbs. per sq. in.

Minimum (20 ft. ground line) 6·2 lbs. per sq. in.

machine normale.

Poids.—28,420 kilogrammes.

Proportion 248 kilos par cheval.

Pression par centimètre carré.

Maximum 1·70.

Minimum ·43.

WILKINS'S CLUTCH GEAR.

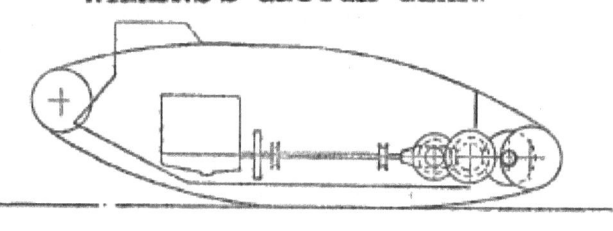

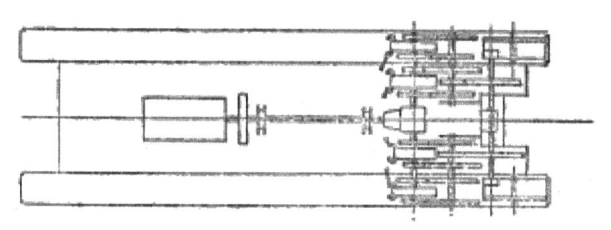

Wilkins's Clutch Gear System Machine, No. 7 (E.M.H.)

Description.—This system of gearing is designed with the object of providing a better and simpler control both for steering and for changing gear.

Power Unit.—One six-cylinder 105 B.h.p. Daimler engine running at 1000 r.p.m.

Transmission.—The drive is taken from the engine by a bevel gear to a cross shaft; at each end of this cross shaft is mounted a three-speed and reverse gear box in which the gears are engaged by means of clutches so constructed that it is impossible to have two gears engaged at the same time. Thence the drive is taken by spur gearing to a divided cross shaft, on which is mounted a pinion engaging with internally cut teeth in the track-driving sprocket. A differential lock is obtained by means of sliding dog clutches on the divided cross shaft.

Speed Control.—This transmission provides three forward speeds and one reverse, any of which may be engaged

Système Wilkins, à embrayge.

Ce système donne un contrôle supérieur et en meme temps plus simple, tant pour la direction que pour le changement de vitesse. Moteur Daimler 105 chevaux.

Transmission.—La transmission est à roues coniques, avec axe transversal, à chaque bout duquel se trouve un engrenage à trois vitesses et marche en arrière, dont les roues sont engagées par des embrayages à plaques, de manière qu'il est impossible que deux embrayages soient engagés en même temps. Dans cet appareil l'effort est transmis par un autre axe transversal sur lequel est monté un pignon, qui s'engage avec la roue intérieurement dentée de la chenille.

L'action différentielle est obtenue à volonté par le déplacement d'un manchon d'accouplement monté sur les deux parties de l'axe qui passe à travers la machine. Il y a trois vitesses en avant, et une en arrière.

Direction.—La direction se fait par l'engagement de l'em-

whilst the machine is in motion.

Steering.—The steering on this machine is carried out by engaging a lower gear in the gear box on the side to which it is desired to turn.

Estimated Speeds.—
1st gear, 1·0 mile per hour.
2nd ,, 1·9 miles per hour.
3rd ,, 3·9 ,, ,,

Armour.—Same as No. 1.
Armament.—Same as No. 1.
Weight.—The weight of this machine complete is about 26 tons.
Weight, horse power ratio, 555 lbs. per h.p.

Track Pressure.—
Maximum (5 ft. ground line) 23·1 lbs. per sq. in.
Minimum (20 ft. ground line) 5·8 lbs. per sq. in.

brayage de la vitesse inférieure d'un côté ou de l'autre, selon la direction voulue.

Blindage.—Armement et poids, et proportion de poids, aussi les pressions de chenilles sont identiques avec ceux de la machine normale.

GUN-CARRYING MACHINE.

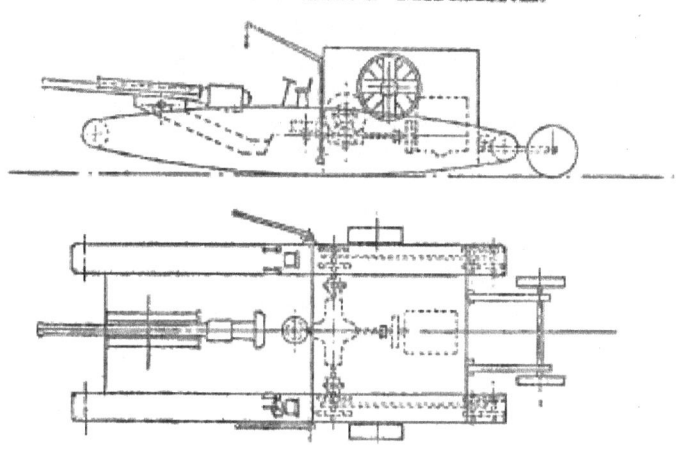

Gun-Carrying Machine No. 8 (G.C.)

Description.—This machine carries either—

(a) One 60-pounder 5-in. gun with carriage, wheels and 64 rounds of ammunition, or

(b) One 6-in. howitzer complete with carriage, wheels and 64 rounds of ammunition.

Either gun can be fired from the machine, or can by a special slipway be unshipped with its carriage and mounted on its wheels. This machine can carry ammunition alone up to 130 rounds of 60-pounder ammunition or 130 rounds of 6-in. howitzer ammunition.

Power Unit.—The power unit on this machine consists of a single six-cylinder 105 B.h.p. Daimler engine running at 1000 r.p.m.

Transmission.—The transmission on this machine is very similar to that used in the Standard Machine No. 1. The chief difference is that the track driving sprockets are wider and stronger, necessitating a corresponding increase in the width of the track links.

Speed Control.—The speed con-

Machine porteur de canons, Système G.C.

Cette machine est capable de transporter ou un canon portant un obus de 27·2 kilos, 125 mm. complet avec affût de canon, roues démontées, et 64 cartouches, ou un obusier de 150 mm. avec 64 cartouches. On peut tirer avec le canon ou avec l'obusier monté en place sur le porteur.

Par le moyen d'un traineau la pièce peut être enlevée et mise sur roues. La pièce enlevée le porteur peut être chargé de 130 obus.

Transmission.—La transmission est très semblable à celle de ce machine No. 1.

La vitesse de marche est celle de la machine normale.

La direction se fait en partie par l'intermédiaire d'une direction ordinaire d'auto, mais les roues directrices sont sur l'arrière de la machine au lieu de sur le devant. C'est le système qu'on a supprimé dans les premières machines "Tanks," mais qu'on a trouvé très nécessaire pour une machine aussi longue que celle-ci.

La direction est aidée par le freinage d'une chenille ou l'autre, selon la direction vou-

trol is the same as the Standard Machine No. 1.

Steering.—The steering on this machine is carried out partially by means of an ordinary steering wheel in front of the driver's seat, which operates the wheels of the tail, which is used on this machine, and partially by the independent use of the secondary gears.

Estimated Speeds.—
1st gear, ¼ of a mile per hour.
2nd ,, 1¼ miles per hour.
3rd ,, 2 ,, ,,
4th ,, 4 ,, ,,

Weight.—The weight of this machine complete with either gun or howitzer and ammunition is 34 tons. The machine alone weighs 25 tons.

Weight, horse power ratio, 725 lbs. per h.p.

Track Pressure.—
Maximum (4 ft. 3 in. ground line) 35·5 lbs. per sq. in.
Minimum (25 ft. ground line) 6·0 lbs. per sq. in.

lue.

Poids.—Chargé canon ou obusier.
—Munitions et personelle 35,000 kilos, net 25,375 kilos.

Proportion—Poids à puissance de cheval 329 kilos.

Pression sur la chenille.
Maximum 2·39 kilos par centimètre carré.
Minimum ·42 kilos par centimètre carré.
Longueur pour le minimum poids 7·62 m.

Appendix 3

STRENGTH AND ESTABLISHMENT OF THE ARTILLERIE D'ASSAUT FROM 1st JANUARY TO THE 31st OCTOBER, 1918

Date.	New Units formed during the month.	Establishment — Officers Combatant	Establishment — Officers N.-C.[1]	Establishment — Other Ranks Combatant	Establishment — Other Ranks N.-C.[1]	Strength — Officers	Strength — Other Ranks	Casualties in action[2] — Officers	Casualties in action[2] — Other Ranks	Percentage — Officers	Percentage — Other Ranks	Remarks.
Existing on 1st Jan.	H.Q. "A.S."; 2 H.Q.s with Army Groups; 3 Parks, 4 Schneider Batts., H.Q.; 3 St. Chamond Batts., H.Q.; 8 St. Chamond Coys.; 5 Rep. and Supply Sections.	349	9	4,142	582	393	4,064	—	—	—	—	[1] Non-combatant personnel is composed of Workshop personnel. [2] Casualties do not include those evacuated for sickness or other reasons.
31st Jan.	2 St. Chamond Coys., 1 St. Chamond Batt. H.Q.; 1 Renault Coy.	383	9	4,607	582	412	4,744	—	—	—	—	
28th Feb.	2 St. Chamond Coys.; 3 Rep. and Supply Sections; 2 Renault Coys.; 1 Batt. (special) Infantry.	438	9	5,961	582	436	5,902	—	—	—	—	
31st Mar.	3 Renault Batt. H.Q.; 6 Renault Coys.; 2 Batts. (special) Infantry.	513 513	9 9	7,738 7,738	582 582	490 530	7,995 8,809	— 7	— 23	— 1·3	— 0·3	
30th April. 31st May.	4 Renault Batts.; 3 Brigade H.Q.	605	10	9,193	776	609	10,461	—	1	—	—	
30th June.	3 Renault Batts.; 2 Group H.Q. in substitution for 2 H.Q. with Army Groups (see above).											
31st July.	4 Renault Batts., 1 Brigade H.Q.	673	30	10,280	1,448	628	11,146	52	432	8·3	3·9	
31st Aug.	4 Renault Batts., 1 Brigade H.Q.	752	30	11,581	1,448	684	12,713	114	745	16·6	5·9	
30th Sept.	4 Renault Batts., 1 Brigade H.Q.	860	30	13,616	1,448	759	13,990	23	106	3·3	1·5	
31st Oct.	H.Q., 1 Group H.Q.; 3 Renault Batts.	963 1,080	30 30	15,557 16,592	1,448 1,448	820 904	15,451 17,119	68 57	484 522	8·2 0·3	3·1 3·0	

Appendix 4

STRENGTH AND ESTABLISHMENT OF THE TANK CORPS FROM 1st JANUARY TO THE 31st OCTOBER, 1918

Date.	New Units formed during the month.	Establishment — Officers Combatant.	Establishment — Officers N.-C.[1]	Establishment — Other Ranks Combatant.	Establishment — Other Ranks N.-C.[1]	Strength Officers.	Strength Other Ranks.	Casualties in action Officers.	Casualties in action Other Ranks.[2]	Percentage Officers.	Percentage Other Ranks.	Remarks.
Existing on 1st Jan. 1918.	H.Q. Tank Corps. 3 Brigades including 10 Batts. Central Workshops. 2 Salvage Coys. 1 Coy. and 2 Sections G.C. Tanks.	993	24	6,577	1,085	895	7,185	—	—	—	—	[1] Non-combatant personnel is comprised of Workshop personnel. [2] Casualties are those incurred in action only.
Formed during January.	1 Brigade H.Q. 2 Batts. Central Stores. 3 Advanced Workshops, No. 2 Coy. G.C. Tanks Completed. Salvage Coys. Renamed Tank Field Coys.											
F. d. Feb.	No. 13 Tank Batt.	1,188	54	7,250	1,709	1,079	7,999	1	3	·01	·04	
,, Mar.	No change.	1,233	54	8,335	1,709	1,162	8,796	—	2	—	·02	
,, Apr.	No. 17 (Armed Car) Tank Batt.	1,233	54	8,335	1,709	1,194	8,546	44	271	3·5	3·2	
,, May.	No. 4 Advanced Workshops. Nos. 1 and 2 Tank Supply Coys.	1,249	58	8,425	1,783	1,340	9,292	99	610	8·2	6·6	
,, June.	No. 14 Tank Batt., Nos. 3, 4 and 5 Supply Coys.	1,289	58	8,860	1,783	1,355	9,873	36	235	2·5	2·4	
,, July.	No. 15 Tank Batt., No. 5 Advanced Workshops, H.Q. Carrier Units.	1,439	58	9,943	1,783	1,509	11,131	9	29	·6	·2	
,, Aug.	No change.	1,533	62	10,540	1,857	1,589	11,747	30	109	1·9	·9	
,, Sept.	No. 16 Tank Batt., Inspection of Tank Machinery, H.Q. 5th Tank Brigade.	1,533	62	10,540	1,857	1,267	10,721	253	1,116	19·6	10·0	
,, Oct.	H.Q. 6th Tank Brigade. No. 18 Tank Batt.	1,628	65	11,145	1,869	1,383	11,315	86	428	6·1	3·7	
		1,728	65	11,750	1,869	1,391	11,064	149	778	11·4	7·0	

Appendix 5

ARTILLERIE D'ASSAUT

A SHORT SKETCH OF THE DEVELOPMENT OF THE FRENCH TANKS

The existence of the French Tank Corps was due to the untiring energy of one man.

On December 1st, 1915, Colonel Estienne, then commanding the 6th French Divisional Artillery, addressed a letter to the commander-in-chief of the French Armies, in which he expressed his firm belief that it was possible to construct an engine of war mechanically propelled and protected by armour which would transport infantry and guns over the battlefields on the Western Front.

This was the result of his work through the year 1915, during which time he had seen Holt tractors in use with British Artillery Units.

On the 12th of December, 1915, he was given an interview at G.Q.G. (French General Headquarters), where he propounded his theories, and on the 20th he visited Paris and discussed mechanical details with the engineers of the Schneider firm.

It was not, however, until the 25th of February, 1916, that the Under Secretary's Department for Artillery and Munitions decided to place an order for 400 armoured vehicles with the Schneider works.

Colonel Estienne, meanwhile, returned to his command, the 3rd Corps Artillery, before Verdun, but kept in touch with the makers unofficially. Here he learned, on or about the 27th of April, 1916, that 400 other armoured vehicles of a different type had been ordered by the Under Secretary's Department. These were the St. Chamond type—a heavier machine, with petrol-electric motive force.

In June of this year the Ministry of Munitions, which had been created meanwhile, decided to have someone to take charge of construction and early organisation. An area for experiment and instruction was formed at Marly-le-Roi in July, and, later, a depot for the reception of stores at Cercottes, both being under the control of the ministry.

On the 30th of September General Estienne was gazetted "*Commandant de l'Artillerie d'Assaut aux Armées*," and also appointed the commander-in-chiefs delegate to the Ministry of Munitions in matters connected with Tanks. He thus became an official connecting link between the armies in the field and the organisation for construction. The name "*Artillerie d'Assaut*," with its contraction "A.S.", came into use at this time, "S." being used instead of a second "A" for the sake of euphony.

In October a training centre within the army areas was established at Champlieu, on the southern edge of the forest of Compiègne, and it was here that the first unit of tanks arrived on the 1st of December, 1916. It consisted of sixteen Schneider tanks. During the succeeding months Schneider and St. Chamond units continued to arrive at irregular intervals, and by April 1917 nine Schneider Companies and one St. Chamond Company were ready for operations. On the 16th of that month they went into action in the ambitious attack over a wide front on the heights above the Aisne. The attack had been postponed for two days, and on the 16th the enemy's artillery was not mastered. Eight Schneider Companies were used. Three companies were to operate between the Craonne Plateau and the Miette, and five companies between the Miette and the Aisne.

The operation was unsuccessful. The former companies failed to get into action, and consequently suffered heavy losses from enemy artillery which overlooked their advance from the heights of Craonne Plateau. The latter companies succeeded in crossing the second line of the enemy's defence and in reaching and even passing the third line, but although they remained for a considerable time in front of the infantry, it was impossible for the infantry to follow them, owing to the very heavy machine-gun fire. At nightfall the tank companies were rallied, having sustained serious losses both in machines and men. Bodies of infantry had been specially detailed to escort tanks and prepare paths for them, but their training with them had been so short that their work was either ineffectual or not done at all. Of 132 tanks, seventy-six remained either ditched or mechanically unfit in or near the enemy's lines. Of these fifty-seven were totally destroyed.

On the 5th of May, one St. Chamond and two Schneider Companies took part in a hurriedly prepared operation with the 6th Army. The Schneider Companies led the infantry in a successful attack on Laffaux Mill. Of the sixteen St. Chamond tanks detailed for the action, one only crossed a German trench.

Between May and October an attack on the west of the Chemin des Dames was carefully prepared by the 6th Army. Special infantry was attached to the Tank Corps as "*troupes d'accompagnement*," and lived with tank units. This was rendered necessary by the inability of the tanks to cross large trenches unaided.

The left of the attack was over the same ground as that of the right in the May attack. Five companies took part in the operation under the orders of Colonel Wahl, who had recently been appointed to command the *Artillerie d'Assaut* with the army. The Schneider Companies again operated with success, while of the St. diamonds only one or two reached the plateau. St. Chamonds were again employed a few days later, but afforded the infantry no support. Of the tanks actually in these two actions (sixty-three), only eight were hit by enemy artillery. All tanks were salved, together with a few lost in the previous engagement. It was considered that the action of the tanks had thoroughly justified their construction.

Meanwhile, General Estienne had been working on the development of a lighter tank. This idea of a light tank first came to him after his visit to England in June 1916, where he had seen the British Mark 1 manoeuvre at Birmingham. His report, written on his return, clearly showed his desire for a heavy tank to have been to a great extent achieved in the machine he had seen in England. His idea of a light tank was the development of an earlier idea of attacking with waves of skirmishers in open order, each skirmisher clad in armour and armed with a rapid-firing weapon. He believed now that the same result might be reached with a light armoured vehicle. These views he laid before the Renault firm in July 1916, and urged the ministry to accept his proposed light Tank, but without success.

Complete designs were, however, prepared, and on the 27th of November, General Estienne was able to propose to the commander-in-chief, Marshal Joffre, the construction of a large number of light tanks for future operations, and to inform him of the existence of plans for such a tank; in fact, 150 had already been ordered as "command" tanks for the heavy battalions. The first trial was held on the 14th of May, 1917. Still the ministry were not convinced, and it was not until further trials had taken place in May that an order for 1,150 tanks was confirmed. In June the number ordered was raised to 3,500. In this month a new sub-department of the Ministry of Munitions was formed to deal specially with tanks, and was called *"La sous Direction d'Artillerie d'Assaut."*

It was the Battle of Cambrai in November 1917 which finally convinced the ministry of the potentialities of the tank. Their opposition ceased, and in order to accelerate the output the firms of Renault, Schneider and Berliet were all engaged in the manufacture of these light tanks, whilst negotiations were opened with America for a further supply of them and an order placed for 1,200.

In December 1917 it was decided to form thirty light tank battalions, of seventy-five tanks each, three of which were to be wireless tanks. For the purposes of secrecy light tank units were organised in an auxiliary park which the ministry established in December within the precincts of the existing centre at Champlieu, but it was not until the beginning of June 1918 that the armies began to receive battalions for operations at the rate of one per week.

Orders for the St. Chamond and Schneider tanks, meanwhile, were limited to 400 of each type. As they became obsolete a heavier type of tank was designed (C.A.3), but only one of these was built. A still heavier type, weighing forty tons, was given a trial in the grounds of the *Forges et Chantiers de la Méditerrannée* in December 1917. It was then decided to build a tank (2.C.) weighing sixty-two tons, which would in all probability reach seventy tons by the time it was finished.

In March 1918, when the German offensive began, all available Tanks were sent up behind the 3rd Army front as counter-attack troops, and in this capacity took part in minor operations for possession of important tactical features, with varied success. Altogether thirty-six tanks were employed in these local operations. Then, on the 27th of May, the Germans launched, between Soissons and Reims, the attack which was aimed at Paris. On the 9th of June it was extended towards the north between Montdidier and Noyon. The attack that day fell on the 3rd Army, behind which four heavy tank battalions were in position. The first and second lines soon gave way, and troops detailed for counter-attack were absorbed in the battle. Reinforcements were hurried up on the 10th, and General Mangin launched his counter-attack with tanks and tired infantry on the 11th. The battle lasted till the 18th. In spite of serious difficulties, 111 out of 144 tanks started at zero hour. Losses in machines were very heavy and casualties in personnel reached 50 *per cent.* Owing to the good going, however, tanks were able to outdistance the infantry and succeeded in inflicting a heavy blow on the enemy. His offensive at this point was definitely broken.

The St. Chamond and Schneider tanks were now becoming rapidly worn out, and as further construction of these types had ceased, their maintenance became more difficult. In the action of June 11th they were at their zenith. From that date onwards they continued to fight until they dropped by the wayside, and gradually heavy units ceased to exist, or were amalgamated with the light. Finally, two battalions were armed with the British Mark V. Star, but these battalions never went into action.

To meet the attack on May 27th, it had been decided to use all available means, and two battalions of Renault tanks were hurried up by road to the north-eastern fringes of the forest of Villers Cotterets from Champlieu, although they were intended originally solely for use in attack. On the 31st of May they first went into action, two companies working with Colonial infantry on the plateau east of Cravançon Farm. From this date to the 15th of June these two battalions continued to act on the defensive with tired infantry. Nevertheless, they succeeded in preventing a further advance of the German Armies.

By the 15th of June reinforcements had arrived, and operations were carried out up to the end of the month which enabled the line to be straightened out and starting-points gained for any future offensive on a large scale. Eighty-five Renault tanks were engaged during the latter part of June.

The next and last attempt of the Germans to break through the Allied line was launched between Château Thierry and Reims on the 15th of July. The attack was awaited with feverish excitement. First the probability of an impending attack in this sector became known, then its boundaries, and finally the actual zero hour. When the attack was launched its weight was wasted on evacuated trenches, and it failed completely. The moment for the counter-attack had arrived, and all available tank units, heavy and light, were hurried, by road and rail, to the west, south-west, and eastern sides of the salient between Soissons and Reims.

The counter-attack was launched on the 18th of July with a view to cutting off the salient, and tank units in varying numbers were employed with each of the three armies engaged.

In the 10th Army area an advance of five to six kilometres was made, with tanks always in the van of the rapidly-tiring infantry. On the first day, out of 324 tanks available 223 were engaged. On the succeeding day this number was reduced to 105. One hundred were again available for action on the 21st. Altogether, during the five days' fighting, 216 Schneider and 131 St. Chamond and 220 Renault tanks fought actions, and of this number 180 were lost, while the casualties in personnel were 819.

In the 6th Army area an advance of twenty kilometres was made during seven days' operation, and forty-three St. Chamond and 230 Renault tanks fought actions. Losses were much less serious owing to the German tendency to retire, fifty-eight tanks only being disabled, while there were seventy-five casualties. The 6th Army was afterwards withdrawn from this front owing to the shortening of the line, and next appears with the Grand Army of Flanders.

In the 5th Army area ninety light tanks fought actions. Among these was a minor operation in conjunction with units of the 22nd British Corps astride the Ardre River.

This operation had tremendous influence on succeeding operations owing to the eagerness with which infantry commanders clamoured for tank units, and the consequent speeding up of training and turning out of new battalions.

From this time new battalions were made available for the forward area at the rate of one per week; thus tired battalions withdrawn on the 23rd and 27th July could almost immediately be replaced.

The British counter-offensive opened on the 8th of August in conjunction with the 1st and 10th French Armies, and on the 8th and 9th of August eighty tanks advanced with the infantry a distance of eighteen kilometres on the south of the Roye-Amiens Road, while thirty tanks made a five-kilometres advance near Montdidier.

An attack on a larger scale was made west of Roye from the 16th to the 18th of July. Here sixty Renault and thirty-two Schneider tanks were engaged. The area of operations was an old battlefield, and tanks found great difficulty in co-operating with the infantry.

The next operation was a continuation of the 10th Army offensive, and took place between the Oise and the Aisne Rivers. It began on the 20th of August, and continued intermittently up to the 3rd of September. On the 20th and 22nd twelve Schneider, twenty-eight St. Chamond and thirty Renault tanks were engaged north of Soissons, and during the week commencing the 28th of August three light battalions advanced five kilometres between the Aisne and the Aillette. Three hundred and five tanks were employed at different times during these operations.

The next operation in which tanks were engaged was the straightening out of the St. Mihiel salient. French tanks were used both with the 2nd French and American Armies. During the two days' fighting, the 12th and 18th of September, twenty-four Schneider, twenty-eight St. Chamond and ninety Renault tanks entered the battlefield.

On the next day the 10th Army resumed its offensive east of Soissons. Eighty-five Renault actions were fought during the three days that it lasted.

Ten days later a larger joint attack was made by the 5th and 2nd French Armies in conjunction with the American Army, and then the 4th Army attacked on a fifteen kilometre front in the Champagne, and employed during the period 680 Renault and twenty-four Schneider tanks.

The attack was very difficult in its initial stages, as it had to be made over highly-organised ground, part of which the French had evacuated in anticipation of the German attack on the 15th of July.

Meanwhile, the 2nd French and American Armies attacked on a twelve kilometre front between the Argonne and the Meuse, and advanced during the seven battle days fifteen kilometres. Two hundred and thirty Renault, thirty-four Schneider and twenty-seven St. Chamond actions were fought during this advance.

At the urgent request of the French Army commander in Flanders, a Renault battalion, less one company, and some heavy units were sent to Dunkirk. The 3rd Company of this Renault battalion had been sent on detachment to join General Franchet d'Esperey at Salonika. On the 30th of September and on the 3rd and 4th of October fifty-five tanks were employed north-west of Roulers, without achieving much success. From the 14th to the 19th of October eight St. Chamond and 170 Renault tank engagements were fought during the offensive and penetrated to a depth of fifteen kilometres, but not a few St. Chamond tanks failed to negotiate the country. The advance was continued on the 31st of the month in the direction of Thielt, and on that and the two succeeding days seventy-five tank engagements took place.

From the end of September onwards operations consisted in following up and pressing upon a retiring enemy all along the line, and small engagements with a few tanks took place in various sectors.

To summarise the number of individual tank engagements during the year 1918: Renault tanks fought 3,140 times, Schneider 473 times, and St. Chamond 375 times, making a total of 3,988 actions. Tanks were employed on 45 of the 120 days between the 15th of July and the 11th of November.

The figures available on the 1st of December, 1918, in regard to tank losses were as follows:—

Renault tanks. Of the total of 2,718 delivered to the fighting Units, 284 remained on the battlefield for salvage, 74 were definitely abandoned, 369 had been returned to the makers for reconstruction, and 1,991 were left available for action with units in the field.

Schneider tanks. Of the 400 manufactured, only 97 were still fit, while 137 existed in more or less unfit condition; the remainder were dead and buried.

St. Chamond tanks. Still fewer remained fit of the original 400. Seventy-two laid claim to be capable of further fighting, 157 were admittedly on the sick list, and the balance had been scrapped.

The German military authorities admitted that their discomfiture on the 18th of July was largely brought about through the use of "masses of tanks," and their *communiqués* were evidence of the great influence that the Renault tank had upon the battle.

The general commanding-in-chief of the French Armies addressed to the *Artillerie d'Assaut*, on the 30th of July, the following order of the day, "*Vous avez bien mérité la Patrie*," while the general commanding received the cravat of the *Legion d'Honneur*, and was promoted to the rank of general of division for his great services to his country.

Appendix 6

TANKS IN PALESTINE

In December 1916, eight tanks (Marks 1 and 2) were sent out to Egypt to join the army operating in Palestine, with twenty-two officers and 226 N.C.O.s and men. No opportunity was given to adapt the tanks, designed for use in France, to the very different conditions of desert warfare, and though only partially successful, they achieved a much greater measure of success than it had seemed possible to hope for them. This was due to the determination and very fine spirit of their crews.

They went into action first in the Second Battle of Gaza, in the middle of April 1917. Owing to the shortness of time there was practically no reconnaissance, and the infantry commanders hardly understood what could and what could not be expected of tanks. As a result the objectives given them were not only very difficult but too many. Eight machines were asked to do what in France would have been given to two battalions. In spite of this the tanks did very valuable work in protecting the infantry, although they were far too few to make the protection anything like adequate, for the Turks were equipped with hundreds of cleverly hidden machine-guns. What the tanks could do they did. Each of the eight covered, on an average, forty miles of country. One was destroyed by a direct hit, and another had one of its tracks broken by a shell and was captured in a Turkish counter-attack.

Seven months later the tanks again went into action in the Third (and victorious) Battle of Gaza. A reinforcement of three Mark 4s had been received, and eight tanks went into action, six in the first line, and two in reserve. They operated with the 54th Division and the Indian cavalry near the sea, and the preliminary reconnaissances were made on horseback and by drifter. The Tanks did useful work, but not all that was asked of them, again for the reason that more was asked than they could possibly have performed. The six first line tanks had no fewer than twenty-nine objectives assigned to them.

The infantry, and not the tanks, began the attack at eleven at night on November 1st. While this attack was in progress the tanks moved up to their starting-points, which they reached by half-past two in the morning, half an hour before their attack was to begin. The moon had then just risen, and it had been hoped that the tanks would have its light to advance by, but a dense haze, thickened by the smoke of the battle, deprived them of this» and they went forward on compass bearings.

Although they did not do all that was expected of them, the tanks materially helped the infantry. Five of the six reached their first objectives, four reached their second, third and fourth, and one reached its fifth also. Three of the six were temporarily out of action before the end of the battle, and so were the two reserve tanks which came up in support an hour after the action began, both loaded with empty sand bags; and these, unfortunately, caught fire. On the other hand, the casualties in personnel were very small, one man killed and two wounded.

The disabled tanks were repaired but they were not used again. What was now wanted for the difficult work of rounding up the rearguard detachments of the retreating Turkish army was a lighter and faster tank. A mission was sent to France to see if a number of Whippets could be obtained, but this mission only reached headquarters in France on March 21st, 1918, the day the great German attack was launched. There was no possibility then of taking any tanks from the Western Front.

The Third Battle of Gaza ended the tank operations in Palestine, The machines there were handed over to the ordnance department at Alexandria and the crews returned to England.

ALSO FROM LEONAUR
AVAILABLE IN SOFTCOVER OR HARDCOVER WITH DUST JACKET

FARAWAY CAMPAIGN by F. James—Experiences of an Indian Army Cavalry Officer in Persia & Russia During the Great War.

REVOLT IN THE DESERT by T. E. Lawrence—An account of the experiences of one remarkable British officer's war from his own perspective.

MACHINE-GUN SQUADRON by A. M. G.—The 20th Machine Gunners from British Yeomanry Regiments in the Middle East Campaign of the First World War.

A GUNNER'S CRUSADE by Antony Bluett—The Campaign in the Desert, Palestine & Syria as Experienced by the Honourable Artillery Company During the Great War.

DESPATCH RIDER by W. H. L. Watson—The Experiences of a British Army Motorcycle Despatch Rider During the Opening Battles of the Great War in Europe.

TIGERS ALONG THE TIGRIS by E. J. Thompson—The Leicestershire Regiment in Mesopotamia During the First World War.

HEARTS & DRAGONS by Charles R. M. F. Crutwell—The 4th Royal Berkshire Regiment in France and Italy During the Great War, 1914-1918.

INFANTRY BRIGADE: 1914 by John Ward—The Diary of a Commander of the 15th Infantry Brigade, 5th Division, British Army, During the Retreat from Mons.

DOING OUR 'BIT' by Ian Hay—Two Classic Accounts of the Men of Kitchener's 'New Army' During the Great War including *The First 100,000 & All In It.*

AN EYE IN THE STORM by Arthur Ruhl—An American War Correspondent's Experiences of the First World War from the Western Front to Gallipoli-and Beyond.

STAND & FALL by Joe Cassells—With the Middlesex Regiment Against the Bolsheviks 1918-19.

RIFLEMAN MACGILL'S WAR by Patrick MacGill—A Soldier of the London Irish During the Great War in Europe including *The Amateur Army, The Red Horizon & The Great Push.*

WITH THE GUNS by C. A. Rose & Hugh Dalton—Two First Hand Accounts of British Gunners at War in Europe During World War 1- Three Years in France with the Guns and With the British Guns in Italy.

THE BUSH WAR DOCTOR by Robert V. Dolbey—The Experiences of a British Army Doctor During the East African Campaign of the First World War.

AVAILABLE ONLINE AT **www.leonaur.com**
AND FROM ALL GOOD BOOK STORES

ALSO FROM LEONAUR
AVAILABLE IN SOFTCOVER OR HARDCOVER WITH DUST JACKET

THE 9TH—THE KING'S (LIVERPOOL REGIMENT) IN THE GREAT WAR 1914 - 1918 *by Enos H. G. Roberts*—Mersey to mud—war and Liverpool men.

THE GAMBARDIER *by Mark Severn*—The experiences of a battery of Heavy artillery on the Western Front during the First World War.

FROM MESSINES TO THIRD YPRES *by Thomas Floyd*—A personal account of the First World War on the Western front by a 2/5th Lancashire Fusilier.

THE IRISH GUARDS IN THE GREAT WAR - VOLUME 1 *by Rudyard Kipling*—Edited and Compiled from Their Diaries and Papers—The First Battalion.

THE IRISH GUARDS IN THE GREAT WAR - VOLUME 1 *by Rudyard Kipling*—Edited and Compiled from Their Diaries and Papers—The Second Battalion.

ARMOURED CARS IN EDEN *by K. Roosevelt*—An American President's son serving in Rolls Royce armoured cars with the British in Mesopatamia & with the American Artillery in France during the First World War.

CHASSEUR OF 1914 *by Marcel Dupont*—Experiences of the twilight of the French Light Cavalry by a young officer during the early battles of the great war in Europe.

TROOP HORSE & TRENCH *by R.A. Lloyd*—The experiences of a British Lifeguardsman of the household cavalry fighting on the western front during the First World War 1914-18.

THE EAST AFRICAN MOUNTED RIFLES *by C.J. Wilson*—Experiences of the campaign in the East African bush during the First World War.

THE LONG PATROL *by George Berrie*—A Novel of Light Horsemen from Gallipoli to the Palestine campaign of the First World War.

THE FIGHTING CAMELIERS *by Frank Reid*—The exploits of the Imperial Camel Corps in the desert and Palestine campaigns of the First World War.

STEEL CHARIOTS IN THE DESERT *by S. C. Rolls*—The first world war experiences of a Rolls Royce armoured car driver with the Duke of Westminster in Libya and in Arabia with T.E. Lawrence.

WITH THE IMPERIAL CAMEL CORPS IN THE GREAT WAR *by Geoffrey Inchbald*—The story of a serving officer with the British 2nd battalion against the Senussi and during the Palestine campaign.

AVAILABLE ONLINE AT **www.leonaur.com**
AND FROM ALL GOOD BOOK STORES

Printed in July 2023
by Rotomail Italia S.p.A., Vignate (MI) - Italy